# IB Maths AA

## Internal Assessment

# IB Maths AA

## Internal Assessment

The Definitive Analysis and Application IA
Guide For the International Baccalaureate Diploma

**Mudassir Mehmood**

**Zouev Elite IB Publishing**

Published 2022

Printed by Zouev Elite IB Publishing

ISBN 978-1-9996115-5-2, paperback.

# TABLE OF CONTENTS

# PART I
## THE MATH AA IA GUIDE

# 1. THE COMPLETE GUIDE FOR THE PERFECT MATH IA

The IA is a short-written report in which the students pick a certain topic of their interest and build on their mathematical concepts. There is no specific difference in an A&A and A&I other than criterion E which mainly lies in the domain of SL and HL respectively. IA is worth 20% of the entire grade. The supervisor is allowed to provide feedback once on the first draft. The final draft is marked by the supervisor and moderated by the IB. IA is marked on a total of 5 criterions.

**Assessment Criteria:**

| | |
|---|---|
| A Presentation | 4 |
| B Mathematical communication | 4 |
| C Personal engagement | 3 |
| D Reflection | 3 |
| E Use of mathematics | 6 |

**How does IA differ from other math Assessments?**

| **Other math Assessment** | **Math IA** |
|---|---|
| • Student has no option to choose a question<br>• There is one correct answer<br>• There is a one or few acceptable methods<br>• There is no need to explain why you are doing what you are doing.<br>• Limited amount of time<br>• Questions will be asked from any topic | • Students choose the topic you want to answer.<br>• There is no single answer to the problem<br>• Students must justify everything what they do in the IA, as marks are awarded based on student's justification.<br>• Unlimited time (Approx. 30 hours)<br>• Should focus on the topic of student's interest. |

## A MUST TO DO EXERCISE BEFORE STARTING THE IA

Prior to start working on your IA, make sure you have a good understanding of the IA components, the assessment criterions and you should have read at least 3 IAs (low scoring, average scoring, and a high scoring). You should evaluate them against the assessment criterions and compare their actual score with the one you awarded against each criterion. Try to figure out where and why is the difference between the two scores.

### Selection of the Topic

Selection of a topic is the trickiest process when it comes to AA SL/HL or AI HL math. One should spend good time on it and should discuss it with the teacher and peers. Selection of topic should be divided into these steps.

### Step 1 – Mind Mapping

Thinking of what stimulate you, it could be something you have been thinking through your childhood, or something you recently have come across. It could be something that interests you, could be your hobby or idea you heard from someone else. Make list of your key words.

Here is an example but the list can go on and on....

| | | |
|---|---|---|
| Physics | Communication | Radiations |
| Chemistry | Internet | Rumor |
| Biology | Games | Computers |
| Geography | Forest | Algorithms |
| Cell phone | Tablet | Logs |
| Signals | Cricket | Water flow |
| Burgers | Football | Sound |
| Fruits | Lottery | Art |
| Light | Euler | Symmetry |
| Patterns | Reiman's sum | Likelihood |
| Correlation | IT | Geometry |
| Models | Businesses | Tessellations |
| Buildings | Virus | Bridges |
| Structure | Bacteria | Paintings |
| Movies | Radioactivity | Orbits |

Start mind mapping using these key words with each key word being the source of generating more thoughts and idea that can be linked together later. Do not focus on just one point instead think freely and develop connections as much as you can.

Start connection these ideas with the five topic/s in math i.e., numbers and patterns, functions, geometry and trigonometry, statistics and probability and calculus. Your key words may come under so many topics in math, do not worry, it is good.

**List of Topics.**

Whenever it comes to take ideas for picking up a topi/question/problem for the IA, I use IB Math resources. Here is the link (https://ibmathsresources.com/maths-ia-maths-exploration-topics/). You can pick any of the topics given below and start your working. I am writing few questions here. There are more than 200 topics/questions available on this website. You can visit this link to access these.

Algebra and number

1) Modular arithmetic – This technique is used throughout Number Theory. For example, Mod 3 means the remainder when dividing by 3.

2) Goldbach's conjecture: "Every even number greater than 2 can be expressed as the sum of two primes." One of the great unsolved problems in mathematics.

3) Probabilistic number theory

4) Applications of complex numbers: The stunning graphics of Mandelbrot and Julia Sets are generated by complex numbers.

**Geometry**

1a) Non-Euclidean geometries: This allows us to "break" the rules of conventional geometry – for example, angles in a triangle no longer add up to 180 degrees. In some geometries triangles add up to more than 180 degrees, in others less than 180 degrees.

1b) The shape of the universe – non-Euclidean Geometry is at the heart of Einstein's theories on General Relativity and essential to understanding the shape and behaviour of the universe.

2) Hexaflexagons: These are origami style shapes that through folding can reveal extra faces.

3) Minimal surfaces and soap bubbles: Soap bubbles assume the minimum possible surface area to contain a given volume.

4) Tesseract – a 4D cube: How we can use maths to imagine higher dimensions.

**Statistics and modelling 1 [topics could be studied in-depth]**

1) Traffic flow: How maths can model traffic on the roads.

2) Logistic function and constrained growth

3) Benford's Law – using statistics to catch criminals by making use of a surprising distribution.

**Statistics and modelling 2 [more simplistic topics: correlation, normal, Chi squared]**

1) grades? Studies have shown that a good night's sleep raises academic attainment.

2) Is there a correlation between height and weight? (pdf). The NHS use a chart to decide what someone should weigh depending on their height. Does this mean that height is a good indicator of weight?

3) Is there a correlation between arm span and foot height? This is also a potential opportunity to discuss the Golden Ratio in nature.

4) Is there a correlation between smoking and lung capacity?

5) Is there a correlation between GDP and life expectancy? Run the Gap minder graph to show the changing relationship between GDP and life expectancy over the past few decades.

## Games and game theory

1) The prisoner's dilemma: The use of game theory in psychology and economics.

2) Sudoku

3) Gambler's fallacy: A good chance to investigate misconceptions in probability and probabilities in gambling. Why does the house always win?

4) Bluffing in Poker: How probability and game theory can be used to explore the the best strategies for bluffing in poker.

## Topology and networks

1) Knots

2) Steiner problem

3) Chinese postman problem – This is a problem from graph theory – how can a postman deliver letters to every house on his streets in the shortest time possible?

4) Travelling salesman problem

## Mathematics and Physics

1) The Monkey and the Hunter – How to Shoot a Monkey – Using Newtonian mathematics to decide where to aim when shooting a monkey in a tree.

2) How to Design a Parachute – looking at the physics behind parachute design to ensure a safe landing!

3) Galileo: Throwing cannonballs off The Leaning Tower of Pisa – Recreating Galileo's classic experiment, and using maths to understand the surprising result.

## Maths and computing

1) The Van Eck Sequence – The Van Eck Sequence is a sequence that we still don't fully understand – we can use programming to help!

2) <u>Solving maths problems using computers</u> – computers are useful in solving mathematical problems. Here are some examples solved using Python.

3) <u>Stacking cannonballs – solving maths with code</u> – how to stack cannonballs in different configurations.

## Further ideas:

1) <u>Radiocarbon dating</u> – understanding radioactive decay allows scientists and historians to accurately work out something's age – whether it be from thousands or even millions of years ago.

2) <u>Gravity, orbits, and escape velocity</u> – Escape velocity is the speed required to break free from a body's gravitational pull. Essential knowledge for future astronauts.

3) <u>Mathematical methods in economics</u> – maths is essential in both business and economics – explore some economics-based maths problems.

## Mind Map

You can start your IA even in year 1, however, the best time to start thinking about your IA is when you are done with at least 80% of your math syllabus as you will have enough of the options available to pick a topic from. All the math topics should be on your tips so that you exactly know what stimulate you should be handled using a particular math topic. Here is the academic mind map for both AI SL/HL and AA SL/HL. Before finalizing your topic of interest, go through these mind maps and make sure you have a solution to your question/problem available in one/few of these topics.

# Application and Interpretation

# Number and Algebra SL

**Number and Algebra**

ERROR AND APPROXIMATION

$$\% \text{age erro} = \left| \frac{V_A - V_E}{V_E} \right| \times 100\%$$

$V_E$ = Exact Value

$V_A$ = Approximate Value

Significant figures

0.002 (1 s.f)

1.002 (4 s.f)

4.0000 (5 s.f)

Decimal places

23.2 (1 d.p)

4.50 (2d.p)

compound interest $\quad FV = PV \left(1 + \frac{r\%}{k}\right)^{nk}$

FINANCE

$\pm FV$ = Future Value

$\pm PV$ = Present value

$r\%$ = interest ( -Ive for depreciation)

$k = C/Y$ ( yearly, quarterly, monthly, weekly, daily).
$\quad k=1 \quad k=4 \quad k=12$
$\quad k=52 \quad k=365$

P/Y  Payment per year  PMT → Payment

Amortization

Ben borrowed a loan of $ 130,000 from bank which would be returned in 25 years with an interest rate at 7% compounded monthly. What will be the monthly installment?

$N = 12 \times 25 = 300 \text{ months}$

$I = 7\%$

$PV = 130,000$

$FV = 0$

$PMT = ? \qquad \$ -919$

$P/Y = 12$

$C/Y = 12$

15

## Functions

input → process → output
$x^2$   $f(x)=y$

**LINEAR FUNCTION**
y-axis
$ax+by+c=0$
$y=mx+c$
x-axis   $y-y_1=m(x-x_1)$

$y=2$
$(0,0.67)$
$x=3/2$
$(0.5,0)$

$f(x)=ax^2+bx+c$ (standard form)

$a>0$   $a<0$   **QUADRATICS**

**DISCRIMINANT**
$\Delta = b^2-4ac$

$\Delta<0$
no real root

$\Delta=0$
Repeated root

$\Delta>0$
Two distinct
Real
roots

$f(x)=a(x-h)^2+k$
**VERTEX FORM**
$(h,k)$

y-intercept
$x=0$

x-intercepts
$y=0$

$(p,0)$   $(0,c)$   $(q,0)$

**axis of symmetry**
$x=-b/2a$

$f(x)=a(x-p)(x-q)$
**FACTOR FORM**

**RATIONAL FUNCTIONS**
$f(x)=\dfrac{4x-2}{2x-3}$

Horizontal asymptote
as $x\to 0$ then $y=2$

vertical asymptote
$2x-3=0$
$2x=3$
$x=3/2$

**EXPONENTIAL AND LOGARITHMIC FUNCTIONS**

$y=a^x$   $y=x$

$y=a^x$   inverses   $\log_a x$

$y=\log_a x$

**COMPOSITE FUNCTION**
$x \to g(x) \to f(g(x))$
$g$   $f$

$f\circ g\circ x=f(g(x))$

$\begin{pmatrix}2\\3\end{pmatrix}\to\begin{pmatrix}4\\9\end{pmatrix}$ one to one

$\begin{pmatrix}2\\-2\end{pmatrix}\to\begin{pmatrix}4\end{pmatrix}$ Many to one

$f\circ f^{-1}(x)=f^{-1}\circ f(x)=x$

$a f(x)$

**TRANSFORMATION**
$-f(x)$
$f(x-h)+k$

## Geometry and Trigonometry

angle of elevation
angle of depression

L
M
$(6,10)$
$A(2,8)$

**slope-intercept form**
$y=mx+C$
$y=\dfrac{1}{2}x+C$
$10=\dfrac{1}{2}\times 6+C$
$10=3+C$
$C=7$
$y=\dfrac{1}{2}x+7$

$m=\dfrac{10-8}{6-2}=\dfrac{2}{4}=\dfrac{1}{2}$

$M=\left(\dfrac{6+2}{2},\dfrac{10+8}{2}\right)$

$=(4,9)$

Gradient of L (perp bisector)

$-\dfrac{1}{\frac{1}{2}}=-2$

Toxic waste problem

Minor arc

**arc length**
$l=\dfrac{\theta}{360}\times 2\pi r$

major arc

**Area**
of sector $=\dfrac{\theta}{360}\times \pi r^2$

**VORONOI DIAGRAM**

Edges
sites
vertices
cells

B
A
F C
E
G D H

16

## Statistics and Probability

| Weekly Study Hours | Math Score % |
|---|---|
| 8 | 62 |
| 12 | 75 |
| 5 | 50 |
| 15 | 80 |
| 16 | 75 |
| 18 | 88 |
| 5 | 80 |

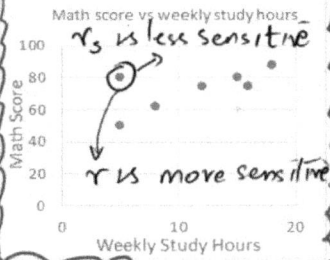

Math score vs weekly study hours

$r_s$ is less sensitive

$r$ is more sensitive

| Rank Hours | Rank Score |
|---|---|
| 5 | 6 |
| 4 | 4.5 |
| 6.5 | 7 |
| 3 | 2.5 |
| 2 | 4.5 |
| 1 | 1 |
| 6.5 | 2.5 |

$\chi^2$ - test GOF

$H_0 : U_1 = U_2$

$Ha : U_1 \neq U_2$ or $U_1 > U_2$ or $U_1 < U_2$

$P > \alpha$ accept $H_0$

$P < \alpha$ Reject $H_0$

$PPMMC = r = 0.669$

$SMRC = r_s = 0.599$

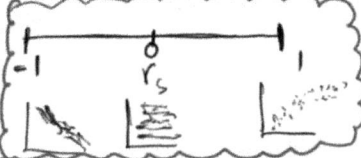

For equal values

$\dfrac{6^{th} + 7^{th}}{2} = 6.5$

$\chi^2$ - test for independence

$d.f = (\text{no of rows} - 1) \times (\text{no of column} - 1)$

$\chi^2 > C.V$ Reject $H_0$

8% → one tailed

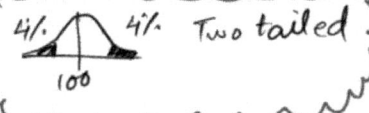
4%   4%   Two tailed

$H_0$ = Null Hyp
$H_a$ = Alternate Hyp
Significance level e.g 5%.

HYPOTHESIS TESTING

---

## Calculus

$\dfrac{d}{dx}(C) = 0$

$\dfrac{d}{dx}(cx) = C$

$\dfrac{d}{dx}(x) = 1$

$\dfrac{d}{dx} x^n = n x^{n-1}$

e.g $y = 4x^3 + x^2 + x - 7$

$\dfrac{dy}{dx} = \dfrac{d}{dx}(4x^3) + \dfrac{d}{dx}x^2 + \dfrac{d}{dx}x - \dfrac{d}{dx}(7)$

$= 4(3x^2) + 2x + 1 - 0$

$= 12x^2 + 2x + 1$

$A \approx \dfrac{h}{2}\left[ (y_0 + y_n) + 2(y_1 + y_2 + \cdots y_{n-1})\right]$

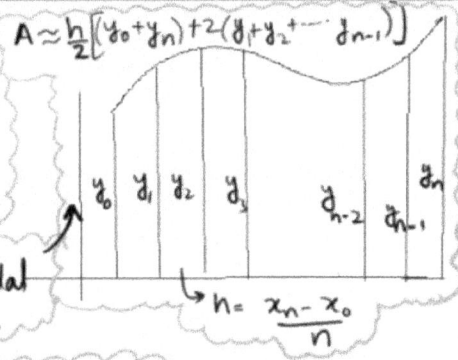

Trapezoidal Rule

$h = \dfrac{x_n - x_0}{n}$

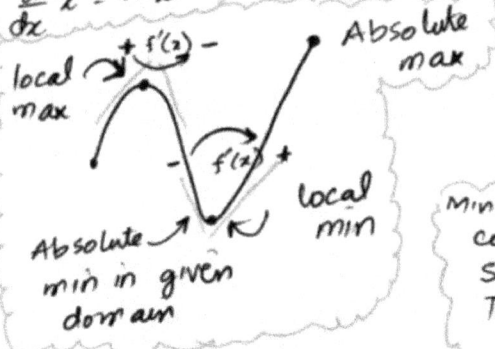

local max

Absolute max

$f'(x)$

$f(x)$

local min

Absolute min in given domain

where $\dfrac{dy}{dx} = 0$

Minimizing
cost
S.A
Time

Optimization

Maximizing
profit
Volume
Sales

e.g

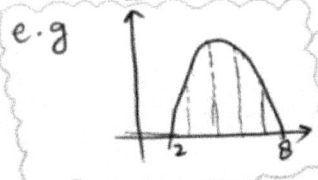

$h = \dfrac{8-2}{5} = 1.2$

$\displaystyle\int_2^8 y\,dx \approx \dfrac{1}{2}(1.2)\left[(0+0) + 2(17.28 + 25.92 + 25.92 + 17.28)\right]$

$= 104 \ unit^2$

| $x$ | 2 | 3.2 | 4.4 | 5.6 | 6.8 | 8 |
|---|---|---|---|---|---|---|
| $y = f(x)$ | 0 | 17.28 | 25.92 | 25.92 | 17.28 | 0 |

**Application and Interpretation (HL only)**

## Geometry and Trigonometry

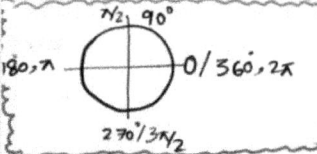

$$2\sin x - \sqrt{3} = 0 \quad 0 \le x \le 2\pi$$
$$\sin x = \sqrt{3}/2$$

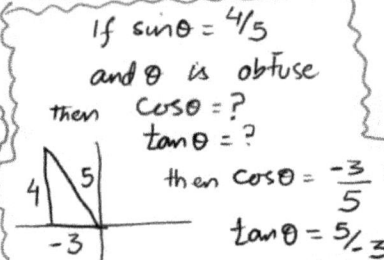

$$x = \pi/6 , 5\pi/6 \quad T \mid C$$

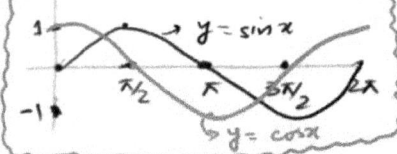

$l = r\theta$   if $r = l$   → 1 radian

$\boxed{l = r\theta}$   $\boxed{A = \frac{1}{2} r^2 \theta}$

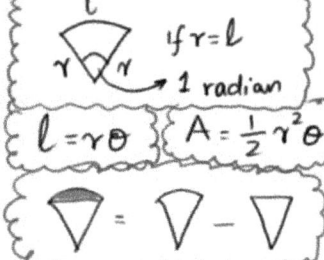

If $\sin\theta = 4/5$
and $\theta$ is obtuse
then $\cos\theta = ?$
$\tan\theta = ?$

then $\cos\theta = \dfrac{-3}{5}$

$\tan\theta = 5/-3$

Degree → Radian

$70° \rightsquigarrow 70 \times \dfrac{\hat{\pi}}{180°}$

Radian → Degree

$3.4 \rightsquigarrow 3.4 \times \dfrac{180}{\pi}$

$m = \tan\theta = y = \tan\theta \cdot x + C$

**TRIG - IDENTITIES**

$\tan\theta = \sin\theta/\cos\theta$

$\sin^2\theta + \cos^2\theta = 1$

$\sin 2\theta = 2\sin\theta\cos\theta$

$\cos 2\theta = \cos^2\theta - \sin^2\theta$
$\quad = 2\cos^2\theta - 1$
$\quad = 1 - 2\sin^2\theta$

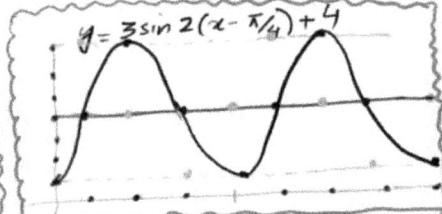

$\sin 4\theta = 2\sin 2\theta \cos 2\theta$

$\sin 8\theta = 2\sin 4\theta \cos 4\theta$

$\cos 4\theta = \cos^2(2\theta) - \sin^2(2\theta)$

---

## Geomtery and Trigonometry (only HL)

**Transition matrix**

$$T = \begin{bmatrix} 0.75 & 0.15 \\ 0.25 & 0.85 \end{bmatrix}$$

Market shares ( example from Revision Village)

Market share after 5 years

$$T^5 S_0 = \begin{bmatrix} 0.75 & 0.15 \\ 0.25 & 0.85 \end{bmatrix}^5 \begin{bmatrix} 1 \\ 0 \end{bmatrix}$$

$$= \begin{bmatrix} 0.432 \\ 0.5764 \end{bmatrix} \text{ by GDC}$$

Reflection in line
$y = (\tan\alpha)x$

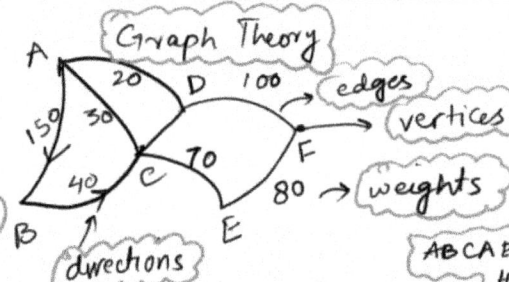

**Graph Theory**

edges
vertices
weights
directions

**Hamiltonian Path**

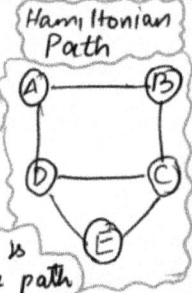

ABCAE is the path

**Minimum Spanning tree**

Prim's or Kruskals Algorithm

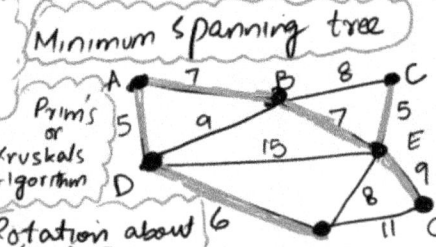

Rotation about $(0,0)$ ↺

$$\begin{bmatrix} a & b \\ c & d \end{bmatrix} = \begin{bmatrix} \cos\theta & -\sin\theta \\ \sin\theta & \cos\theta \end{bmatrix}$$

**Geometric Transformation**

$$\begin{pmatrix} a & b \\ c & d \end{pmatrix}\begin{pmatrix} x \\ y \end{pmatrix} + \begin{pmatrix} h \\ k \end{pmatrix} = \begin{pmatrix} x' \\ y' \end{pmatrix}$$

$$\begin{bmatrix} a & b \\ c & d \end{bmatrix} = \begin{bmatrix} \cos 2\alpha & \sin 2\alpha \\ \sin 2\alpha & -\cos 2\alpha \end{bmatrix}$$

**Euterian cycle**

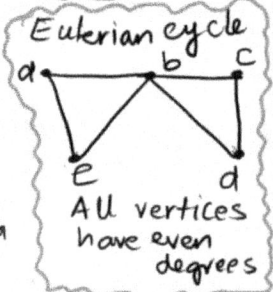

All vertices have even degrees

**Translation** By vector

$$\begin{pmatrix} h \\ k \end{pmatrix}$$

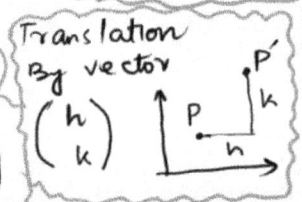

18

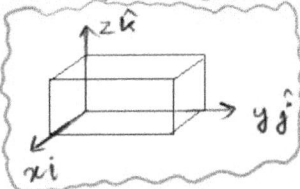

$$U = x\hat{i} + y\hat{j} + z\hat{k}$$

$$|U| = \sqrt{x^2 + y^2 + z^2}$$

$$\hat{U} = \frac{U}{|U|}$$

Vector Product

$$U = \begin{pmatrix} 4 \\ 5 \\ -4 \end{pmatrix} \qquad V = \begin{pmatrix} 3 \\ 2 \\ 1 \end{pmatrix}$$

$$U \times V = \begin{vmatrix} i & j & k \\ 4 & 5 & -4 \\ 3 & 2 & 1 \end{vmatrix}$$

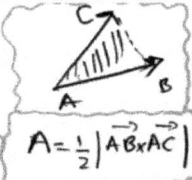

$$A = \frac{1}{2} \left| \overrightarrow{AB} \times \overrightarrow{AC} \right|$$

parallelogram law of addition of vectors

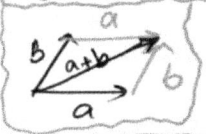

Scalar product

$$U = \begin{pmatrix} U_1 \\ U_2 \\ U_3 \end{pmatrix} \qquad V = \begin{pmatrix} V_1 \\ V_2 \\ V_3 \end{pmatrix}$$

$$U \cdot V = \begin{pmatrix} U_1 V_1 \\ U_2 V_2 \\ U_3 V_3 \end{pmatrix}$$

$$U \cdot V = |U| \times |V| \times \cos\theta$$

$$\therefore \cos\theta = \frac{U \cdot V}{|U| \times |V|}$$

Vector equation of line

$$r = a + tb$$

Position vector → ← direction vector

$$U = 2\hat{i} + 4\hat{j} - 5\hat{k}$$

$$V = 3\hat{i} - 7\hat{j} - 8\hat{k}$$

$$U \cdot V = \begin{bmatrix} 2 \\ 4 \\ -5 \end{bmatrix} \cdot \begin{bmatrix} 3 \\ -7 \\ -8 \end{bmatrix} = 6 - 28 + 40 = 18$$

If $U \cdot V > 0$ acute angle

$U \cdot V = 0$ Right angle

$U \cdot V < 0$ Obtuse angle

---

$$\bar{x} - z\frac{\delta}{\sqrt{n}} < \mu < \bar{x} + z\frac{\delta}{\sqrt{n}}$$

confidence of Interval

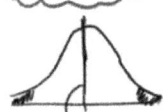

90%

$\delta$ is known

$$\bar{x} - t\frac{S_{n-1}}{\sqrt{n}} < \mu < \bar{x} + t\frac{S_{n-1}}{\sqrt{n}}$$

$df = 12$

95%

$\delta$ is unknown

Quadratic     Sine

Non-linear Regression by GDC

cubic     Exponential logarithmic

$$X \sim N(\mu, \delta^2)$$

$\delta$ is unknown

$$t_{n-1} = \frac{\bar{x} - \mu}{\frac{S_{n-1}}{\sqrt{n}}}$$

$\delta$ is known

$$Z = \frac{\bar{X} - \mu}{\frac{\delta}{\sqrt{n}}}$$

Poisson Distribution

$X \sim P_0(m)$

$$P(X = x) = \frac{m^x e^{-m}}{x!}, \quad x = 0, 1, 2, 3 \cdots$$

$E(X) = m$

$Var(X) = m$

$X \sim P_0(m_1) \qquad Y \sim P_0(m_2)$

$X + Y \sim P_0(m_1 + m_2)$

Explains proportion of variability in y w.r.t 'x'

coefficient of Determination

$$R^2 = 1 - \frac{SS_{RES}}{SS_{TOT}}$$

$$R^2 = 1 \Leftrightarrow SS_{RES} = 0$$

$$\frac{dy}{dx} = \lim_{h \to 0} \frac{f(x+h) - f(x)}{h}$$

$$\frac{dy}{dx} = y' = f'(x)$$

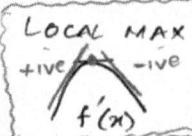 LOCAL MAX
+ive ⟋⟍ -ive
/ $f'(x)$

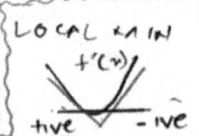 LOCAL MIN
$f'(x)$
+ive — -ive

Graphical Interpretation of $y \to y' \to y''$

$$\frac{d}{dx}(C) = 0$$

$$\frac{d}{dx}(Cx) = C$$

$$\frac{d}{dx} x^n = nx^{n-1}$$

$$\frac{d}{dx}(UV) = UV' + VU'$$

$$\frac{d}{dV}\left(\frac{U}{V}\right) = \frac{VU' - UV'}{V^2}$$

$$\frac{d}{dx} \sin x = \cos x$$

$$\frac{d}{dx} \cos x = -\sin x$$

$$\frac{d}{dx} \ln|x| = \frac{1}{x}$$

$$\frac{d}{dx} e^x = e^x$$

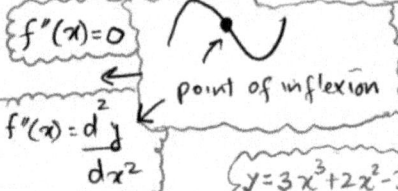 $f''(x) = 0$

point of inflexion

$$f''(x) = \frac{d^2 y}{dx^2}$$

$y = 3x^3 + 2x^2 - 3x - 1$
$y' = 9x^2 + 4x - 3$
$y'' = 18x + 4$

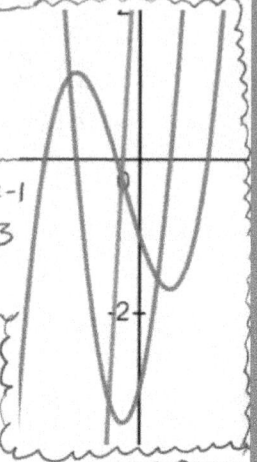
-2-

**CHAIN RULE**

$$\frac{dy}{dx} = \frac{dy}{du} \times \frac{du}{dx}$$

Example $y = \ln \ln|x|$

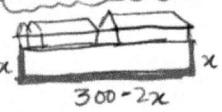 **OPTIMIZATION**

300 m rope
Area to max?
$A = x(300 - 2x)$
$A = 300x - 2x^2$
$A'(x) = 300 - 4x$
$A'(x) = 0 \Rightarrow x = 75$
$A(75) = 11250 \text{ m}^2$

x ⌂ x
$300 - 2x$

$\Rightarrow$ $y = \ln(\ln|x|)$
let $U = \ln x$ then $y = \ln U$
$\frac{du}{dx} = \frac{1}{x}$  $\frac{dy}{du} = \frac{1}{U}$

$\therefore$ $\frac{dy}{dx} = \frac{dy}{du} \times \frac{du}{dx}$
$= \frac{1}{U} \times \frac{1}{x}$
$= \frac{1}{\ln x} \times \frac{1}{x}$

---

$$\int 1 \, dx = x + c$$

$$\int x^n \, dx = \frac{x^{n+1}}{n+1} + c$$

$$\int \cos x \, dx = \sin x + c$$

$$\int \sin x \, dx = -\cos x + c$$

$$\int e^x = e^x + c$$

$$\int \frac{1}{x} \, dx = \ln|x| + c$$

Area Under Curve

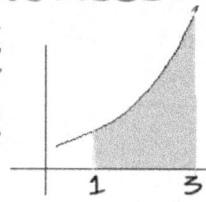
$y = \frac{1}{2}x^2 + 2$
$\int_1^3 (\frac{1}{2}x^2 + 2) \, dx$
$\left[\frac{1}{2}\frac{x^3}{3} + 2x\right]_1^3 = 8.33 \text{ unit}^2$

1    3

**KINEMATICS**
$S = \text{displacement}$ ↑$\int dt$
$\frac{ds}{dt}$ ↓ $V = \text{velocity}$ ↑ $\int dt$
$\frac{dv}{dt}$ ↓ $a = \text{acceleration}$

**FURTHER INTEGRATION**

$$\int \frac{f'(x)}{f(x)} \, dx = \ln|f(x)| + c \Rightarrow$$

$$\int \frac{1}{x \ln x} \, dx \Rightarrow \int \frac{\frac{1}{x}}{\ln x} \, dx$$

$$= \ln \ln|x| + c$$

**BY SUBSTITUTION**

$$\int x(2x^2 + 8)^8 \, dx \Rightarrow$$

$U = 2x^2 + 8$
$du = 4x \, dx$
$\frac{du}{4} = x \, dx$

$$\int U^8 \cdot \frac{du}{4} = \frac{1}{4} \int U^8 \, du$$

$$= \frac{1}{4} \frac{U^{8+1}}{(8+1)} = \frac{1}{36}(2x^2 + 8)^9 + c$$

$$\int x^2 e^{4x^3 + 3} \, dx = \frac{1}{12}\int e^U \, du$$

$U = 4x^3 + 3$
$du = 12x^2 \, dx$
$\frac{du}{12} = x^2 \, dx$

$= \frac{1}{12} e^U$

$= \frac{1}{12} e^{4x^3 + 3} + c$

## Calculus

**Total Shaded Area** $= \int_0^1 f(x)dx + \int_1^3 f(x)dx - \int_3^{3.5} f(x)dx$

### Volume of Revolution

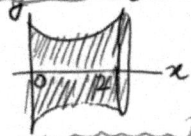

**360° or $2\pi$ about x-axis**

$$V = \int_a^b \pi \left[ f(x) \right]^2 dx$$

$f(x) = 3x^3 + 2x^2 - 3x + 4$

$$\text{Volume} = \pi \int_0^2 (3x^3 + 2x^2 - 3x + 4)^2 dx$$

$$= 947 \ \text{unit}^3$$

**360° or $2\pi$ about x-axis**

$$V = \int_a^b \pi \, x^2 \, dy$$

$\dfrac{dy}{dx} = f(x, y)$  First Order

**Second order**
↓
**Euler's Method**   **Differential Equation**

$$y_{n+1} = y_n + h \times f(x_n, y_n) ;$$
$$x_{n+1} = x_n + h \ \text{(step length)}$$

**For coupled Equations**
$$x_{n+1} = x_n + h \times f(x_n, y_n, t_n)$$

$t_{n+1} = t_n + h$
$y_{n+1} = y_n + f_2(x_n, y_n, t_n)$

**SLOPE FIELD**

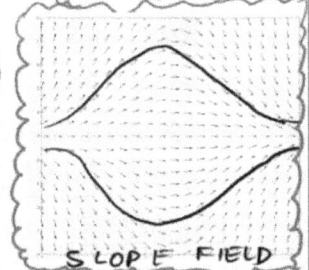

## Analysis and Approaches (SL)

### Number and Algebra

$S_\infty = 80 + 40 + 20 + 10 + 5 + \cdots = \dfrac{80}{1 - 0.5} = 160$
$u_1 = 80, \ r = 0.5$

**INFINITE GEOMETRIC SEQUENCE**
when $0 < |r| < 1$
$S_\infty = u_1 + u_1 r + u_1 r^2 + u_1 r^3 + \cdots = \dfrac{u_1}{1 - r}$

**PASCAL'S △**

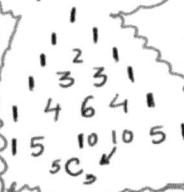

$5C_3$

$^nC_r = \dfrac{n!}{r!(n-r)!}$
where $n! = n \times (n-1) \times (n-2) \times \cdots 3 \cdot 2 \cdot 1$

**BINOMIAL THEOREM**

**FINDING $n^{th}$ Term**
$\binom{n}{r} a^{n-r} b^r$

$(a+b)^n = \ ^nC_0 a^n b^0 + \ ^nC_1 a^{n-1} b^1 + \ ^nC_2 a^{n-2} b^2 + \cdots \ ^nC_n a^0 b^n$

### CAWS OF LOGARITHM

$\log_a xy = \log_a x + \log_a y$
$\log_a \left( \dfrac{x}{y} \right) = \log_a x - \log_a y$
$\log_a x^n = n \log_a x$
$7^{\log_7 100} = 100$

### LAWS OF EXPONENTS / RATIONAL EXPONENTS

$a^m \cdot a^n = a^{m+n}$
$\dfrac{a^m}{a^n} = a^{m-n}$
$(a^m)^n = a^{mn}$

$a^{1/m} = \sqrt[m]{a}$
$a^{m/n} = \sqrt[n]{a^m}$ or
$a^{n/m} = \sqrt[m]{a^n}$
$a^{-m} = \dfrac{1}{a^m}$

**CHANGE OF BASE RULE**

$\log_{10} 100 = \dfrac{\log_2 100}{\log_2 10} = \dfrac{\log_5 100}{\log_5 10}$

$\log_2 (32) = \log_2 2^5 = 5 \log_2 2 = 5$
$\log_2 (14) = \log_2 7 + \log_2 2$
$\log_2 \left( \dfrac{12}{5} \right) = \log_2 12 - \log_2 5$

## Functions

input → process → output
$x^2$    $f(x)=y$

**LINEAR FUNCTION**
$y=mx+c$
$ax+by+c=0$
$y=mx+c$
$y-y_1 = m(x-x_1)$

$f(x)=ax^2+bx+c$ (standard form)

$a>0$

$a<0$

**QUADRATICS**

$f(x)=a(x-h)^2+k$
**VERTEX FORM**
$(h,k)$

y-intercept
$x=0$

x-intercepts
$y=0$

$(p,0)$   $(q,0)$
$(0,c)$

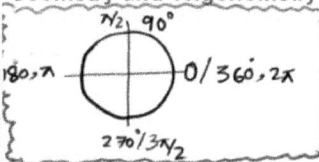

$f(x)=a(x-p)(x-q)$
**FACTOR FORM**

**DISCRIMINANT**
$\Delta = b^2 - 4ac$

$\Delta < 0$
no real root

$\Delta = 0$
Repeated root

$\Delta > 0$
Two distinct
Real
roots

axis of
symmetry
$x = -b/2a$

$y=2$
$(0,0.67)$
$x=3/2$
$(0.5,0)$

**RATIONAL FUNCTIONS**
$f(x) = \dfrac{4x-2}{2x-3}$

Horizontal asymptote
as $x \to 0$ then $y = 2$

vertical asymptote
$2x-3=0$
$2x=3$
$x=3/2$

**EXPONENTIAL AND LOGARITHMIC FUNCTIONS**

$y=a^x$

$y=x$

$y = a^x$
$\log x$   Inverses
$y = \log_a x$

**COMPOSITE FUNCTION**

$x \to g(x) \to f(g(x))$
   $g$      $f$

$fogox = f(g(x))$

$2 \to 4$   one to one
$3 \to 9$

$2 \to 4$   Many to one
$-2$

$fof^{-1}(x) = f^{-1}of(x)=x$

$af(x)$
   $a$

**TRANSFORMATION**
$-f(x)$

$f(x-h)+k$

---

## Geometry and Trigonometry

$\pi/2$, $90°$
$180, \pi$ — $0/360, 2\pi$
$270°/3\pi/2$

if $r=l$
1 radian

$l = r\theta$    $A = \frac{1}{2}r^2\theta$

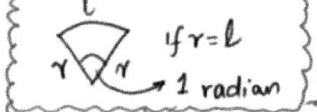 = $\bigvee$ − $\bigvee$

Degree → Radian
$70° \to 70 \times \dfrac{\pi}{180°}$

Radian → Degree
$3.4 \to 3.4 \times \dfrac{180}{\pi}$

$2\sin x - \sqrt{3} = 0$    $0 \le x \le 2\pi$
$\sin x = \sqrt{3}/2$
   S | A
   T | C
$x = \pi/6, 5\pi/6$

If $\sin\theta = 4/5$
and $\theta$ is obtuse
then $\cos\theta = ?$
$\tan\theta = ?$
then $\cos\theta = \dfrac{-3}{5}$
$\tan\theta = 5/-3$

$m = \tan\theta = y = \tan\theta \cdot x + C$

**TRIG - IDENTITIES**

$\tan\theta = \sin\theta/\cos\theta$

$\sin^2\theta + \cos^2\theta = 1$

$\sin 2\theta = 2\sin\theta\cos\theta$
$\cos 2\theta = \cos^2\theta - \sin^2\theta$
$= 2\cos^2\theta - 1$
$= 1 - 2\sin^2\theta$

$y = \sin x$
$\pi/2$   $\pi$   $3\pi/2$   $2\pi$
$y = \cos x$

$y = 3\sin 2(x - \pi/4) + 4$

$\sin 4\theta = 2\sin 2\theta \cos 2\theta$
$\sin 8\theta = 2\sin 4\theta \cos 4\theta$
$\cos 4\theta = \cos^2(2\theta) - \sin^2(2\theta)$

### Mutually Exclusive Event

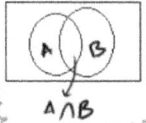

$$P(A \cup B) = P(A) + P(B)$$

### Combined Events

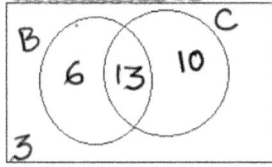

$A \cap B$

$$P(A \cup B) = P(A) + P(B) - P(A \cap B)$$

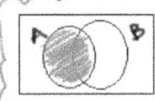
→ $A \cap B'$

→ $A' \cap B$

32 students
19 takes BIO
23 Chem
3 Neither

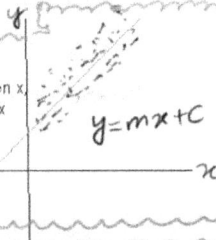

B    C
6  13  10
3

$32 - 3 = 29 \neq 23 + 19$

**Linear Regression**

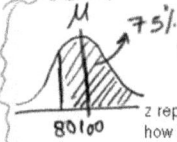

Can be used to predict y for given x, cannot be used x for given y

$y = mx + c$

$E(X) = np \quad V(X) = np(1-p)$

CONDITIONAL PROBABILITY

$$P(A/B) = \frac{P(A \cap B)}{P(B)}$$

$$P(C/B') = \frac{10}{13}$$

If C and B are independent?

$$P(C) \cdot P(B) = P(C \cap B)$$

$$\frac{23}{32} \times \frac{19}{32} \neq \frac{13}{32}$$

So they are independent.

75%

STANDARDISED NORMAL VARIABLE

z repesents how many standard deviation X is away from its mean

80 100

$\sigma = ?$

$-0.674 = \frac{80-100}{\sigma}$

$Z \sim N(0,1)$

$Z = \frac{X - \mu}{\sigma}$

$\sigma = 29.7$

---

## Calculus

$$\frac{dy}{dx} = \lim_{h \to 0} \frac{f(x+h) - f(x)}{h}$$

$$\frac{dy}{dx} = y' = f'(x)$$

$$\frac{d}{dx}(C) = 0$$

$$\frac{d}{dx}(Cx) = C$$

$$\frac{d}{dx} x^n = n x^{n-1}$$

$$\frac{d}{dx}(UV) = UV' + VU'$$

$$\frac{d}{dv}\left(\frac{U}{V}\right) = \frac{VU' - UV'}{V^2}$$

$$\frac{d}{dx} \sin x = \cos x$$

$$\frac{d}{dx} \cos x = -\sin x$$

$$\frac{d}{dx} \ln|x| = \frac{1}{x}$$

$$\frac{d}{dx} e^x = e^x$$

LOCAL MAX
+ive   −ive
$f'(x)$

LOCAL MIN
$f'(x)$
+ive   −ive

$f''(x) = 0$

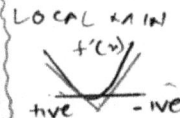
point of inflexion

$$f''(x) = \frac{d^2 y}{dx^2}$$

Graphical Interpretation of $y \to y' \to y''$

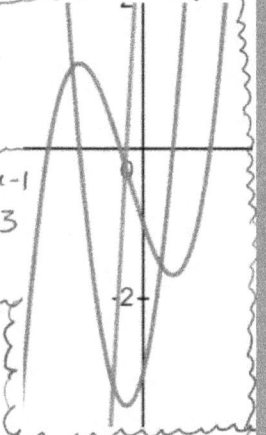

$y = 3x^3 + 2x^2 - 3x - 1$
$y' = 9x^2 + 4x - 3$
$y'' = 18x + 4$

### CHAIN RULE

$$\frac{dy}{dx} = \frac{dy}{du} \times \frac{du}{dx}$$

Example $y = \ln \ln|x|$

### OPTIMIZATION

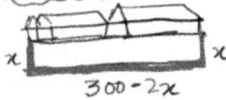
$300 - 2x$

300 m rope
Area to max?
$A = x(300 - 2x)$
$A = 300x - 2x^2$
$A'(x) = 300 - 4x$
$A'(x) = 0 \implies x = 75$
$A(75) = 11250 \text{ m}^2$

$y = \ln(\ln|x|)$

let $u = \ln x$ then $y = \ln u$,

$\frac{du}{dx} = \frac{1}{x}$  $\frac{dy}{du} = \frac{1}{U}$

$\therefore \frac{dy}{dx} = \frac{dy}{du} \times \frac{du}{dx}$

$= \frac{1}{U} \times \frac{1}{x}$

$= \frac{1}{\ln x} \times \frac{1}{x}$

## Calculus

$$\int 1\,dx = x + c \qquad \int \sin x\,dx = -\cos x + c$$

$$\int x^n dx = \frac{x^{n+1}}{n+1} + c \qquad \int e^x = e^x + c$$

$$\int \cos x\,dx = \sin x + c \qquad \int \frac{1}{x}\,dx = \ln|x| + c$$

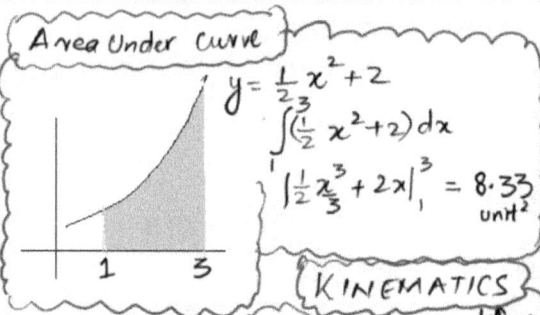

**Area Under Curve**

$$y = \frac{1}{2}x^2 + 2$$

$$\int_1^3 (\frac{1}{2}x^2 + 2)\,dx$$

$$\left| \frac{1}{2}\frac{x^3}{3} + 2x \right|_1^3 = 8.33\ \text{unit}^2$$

### FURTHER INTEGRATION

$$\int \frac{f'(x)}{f(x)}\,dx = \ln|f(x)| + c \implies$$

$$\int \frac{1}{x \ln x}\,dx \implies \int \frac{\frac{1}{x}}{\ln x}\,dx$$

$$= \ln \ln|x| + c$$

### KINEMATICS

$$\frac{ds}{dt} \quad S = \text{displacement} \quad \int dt$$

$$\quad V = \text{velocity} \quad \int dt$$

$$\frac{dv}{dt} \quad a = \text{acceleration}$$

### BY SUBSTITUTION

$$\int x(2x^2 + 8)^8\,dx \implies \quad U = 2x^2 + 8$$
$$du = 4x\,dx$$

$$\int U^8 \cdot \frac{du}{4} = \frac{1}{4}\int U^8 du \qquad \frac{du}{4} = x\,dx$$

$$= \frac{1}{4}\frac{U^{8+1}}{(8+1)} = \frac{1}{36}(2x^2+8)^9 + c$$

$$\int x^2 e^{4x^3 + 3}\,dx = \frac{1}{12}\int e^u du$$

$$U = 4x^3 + 3$$
$$du = 12x^2 dx \qquad = \frac{1}{12}e^u$$
$$\frac{du}{12} = x^2 dx \qquad = \frac{1}{12}e^{4x^3+3} + c$$

---

## Number and Algebra (only HL)

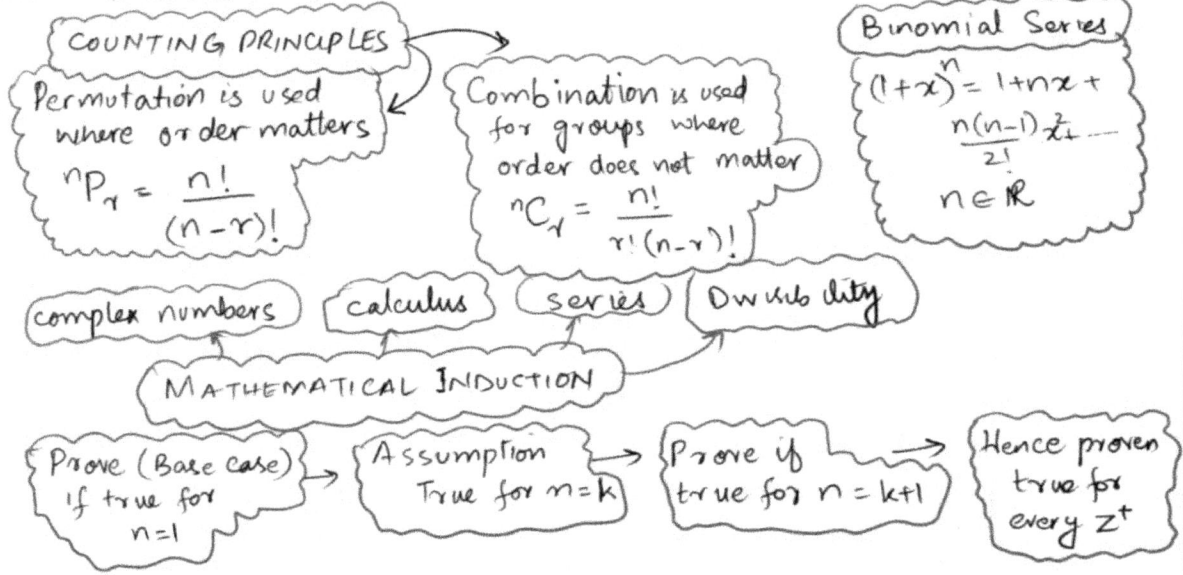

**COUNTING PRINCIPLES**

Permutation is used where order matters

$$^nP_r = \frac{n!}{(n-r)!}$$

Combination is used for groups where order does not matter

$$^nC_r = \frac{n!}{r!(n-r)!}$$

**Binomial Series**

$$(1+x)^n = 1 + nx + \frac{n(n-1)}{2!}x^2 \cdots$$

$$n \in \mathbb{R}$$

complex numbers    calculus    series    Divisibility

**MATHEMATICAL INDUCTION**

Prove (Base case) if true for $n = 1$ → Assumption True for $n = k$ → Prove if true for $n = k+1$ → Hence proven true for every $Z^+$

24

# Analysis and Approaches (HL)

## Number and Algebra (only HL)

**Cartesian Form**

real / imaginary
$$z = a + ib$$
CARTESIAN FORM

$$r^2 = a^2 + b^2$$
$$\theta = \tan^{-1}\left(\frac{b}{a}\right)$$

$$z = r e^{i\theta}$$
Euler's form

**De Moivre's Theorem**
$$(r\,cis\,\theta)^n = r^n\,cis\,n\theta$$

**SYSTEM OF LINEAR EQUATIONS**

**ARGAND DIAGRAM**

$$Mod\ z = |z| = r$$
$$Arg\ z = \theta$$

**COMPLEX NUMBERS**

$$z = r\,cis\,\theta$$
$$z = r(\cos\theta + i\sin\theta)$$
Mod- Argument form
polar form

**Product of $z_1 z_2$**
$$z_1 = 5\,cis\,\frac{\pi}{2} \quad z_2 = 3\,cis\,\frac{\pi}{3}$$
$$z_1 z_2 = 15\,cis\left(\frac{\pi}{2} + \frac{\pi}{3}\right)$$

**Quotient of $\frac{z_1}{z_2}$**
$$= \frac{5\,cis\,\frac{\pi}{2}}{3\,cis\,\frac{\pi}{3}}$$
$$= \frac{5}{3}\,cis\left(\frac{\pi}{2} - \frac{\pi}{3}\right)$$

**nth Root of complex NO**
$$z = 64^{1/3}\left(\cos\pi + i\sin\pi\right)^{1/3}$$
$$= 4\left[\cos\left(\frac{\pi + 2k\pi}{3}\right) + i\sin\left(\frac{\pi + 2k\pi}{3}\right)\right]$$

$$3x + 4y - 9z = 15$$
$$4x - 7y + 10z = 7$$
$$13z + 8y$$

General solution
No Solution
Infinite sol
Unique Sol

Elite IA document - Word    × 1S1stant

## Functions

$$P(x) = x^2 - Sum(x) + Product$$

**POLYNOMIALS**

If $f(a) = 0$ then $(x - a)$ is a factor

2R, 2C

Repeated roots

**Even function**

$$f(-x) = f(x)$$
eg $f(x) = x^2$
$$f(-x) = x^2$$

**odd Function**
$$f(-x) = -f(x)$$
$$f(x) = x^3$$
$$f(-x) = -x^3$$

$$ax^2 + bx + c = 0$$
if $\alpha, \beta$ are two roots
$$\alpha + \beta = -b/a$$
$$\alpha\beta = \frac{c}{a}$$

$$ax^3 + bx^2 + cx + d = 0$$
$$\alpha + \beta + \gamma = \frac{-b}{a}$$
$$\alpha\beta\gamma = \frac{-d}{a}$$

**Vieta's Rule**
$$Sum = \frac{-a_{n-1}}{a_n}$$
$$Product = \frac{(-1)^n a_0}{a_1}$$

**MODULUS**
$$|x| = x \quad if\ x \geqslant 0$$
$$|x| = -x \quad if\ x < 0$$

25

$$\sec x = \frac{1}{\cos x}$$

$$\csc x = \frac{1}{\sin x}$$ Reciprocal Trigonometric Ratios

$$\cot x = \frac{1}{\tan x}$$

$$1 + \tan^2 \theta = \sec^2 \theta$$
$$1 + \cot^2 \theta = \csc^2 \theta$$

COMPOUND ANGLE IDENTITY

$$\sin(\alpha \pm \beta) = \sin\alpha\cos\beta \pm \cos\alpha\sin\beta$$

$$\cos(\alpha \pm \beta) = \cos\alpha\cos\beta \mp \sin\alpha\sin\beta$$

$$\tan(\alpha \pm \beta) = \frac{\tan\alpha \pm \tan\beta}{1 \mp \tan\alpha\tan\beta}$$

$f(x)$ is a continuous at $x = c$ if
$$\lim_{x \to c} f(x) = f(c) \text{ within domain}$$

LIMITS AND CONTINUITY

VERTICAL ASYMPTOTE

$$\lim_{x \to c} f(x) = \infty$$

HORIZONTAL ASYMPTOTE

$$\lim_{x \to \infty} f(x) = k$$

SANDWICH THEOREM

$$\lim_{\theta \to 0} \frac{\sin\theta}{\theta} = 1$$

Geometry and Trigonomtery (only HL)

$$\sec x = \frac{1}{\cos x}$$

$$\csc x = \frac{1}{\sin x}$$ Reciprocal Trigonometric Ratios

$$\cot x = \frac{1}{\tan x}$$

$$1 + \tan^2 \theta = \sec^2 \theta$$
$$1 + \cot^2 \theta = \csc^2 \theta$$

COMPOUND ANGLE IDENTITY

$$\sin(\alpha \pm \beta) = \sin\alpha\cos\beta \pm \cos\alpha\sin\beta$$

$$\cos(\alpha \pm \beta) = \cos\alpha\cos\beta \mp \sin\alpha\sin\beta$$

$$\tan(\alpha \pm \beta) = \frac{\tan\alpha \pm \tan\beta}{1 \mp \tan\alpha\tan\beta}$$

$f(x)$ is a continuous at $x = c$ if
$$\lim_{x \to c} f(x) = f(c) \text{ within domain}$$

LIMITS AND CONTINUITY

VERTICAL ASYMPTOTE

$$\lim_{x \to c} f(x) = \infty$$

HORIZONTAL ASYMPTOTE

$$\lim_{x \to \infty} f(x) = k$$

SANDWICH THEOREM

$$\lim_{\theta \to 0} \frac{\sin\theta}{\theta} = 1$$

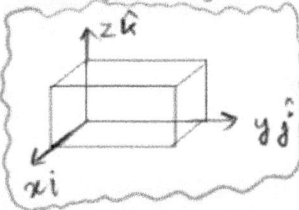

$$U = x\hat{i} + y\hat{j} + z\hat{k}$$

$$|U| = \sqrt{x^2 + y^2 + z^2}$$

$$\hat{U} = \frac{U}{|U|}$$

**Vector Product**

$$U = \begin{pmatrix} 4 \\ 5 \\ -4 \end{pmatrix} \qquad V = \begin{pmatrix} 3 \\ 2 \\ 1 \end{pmatrix}$$

$$U \times V = \begin{vmatrix} i & j & k \\ 4 & 5 & -4 \\ 3 & 2 & 1 \end{vmatrix}$$

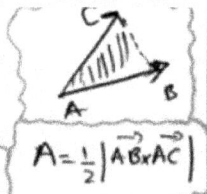

$$A = \tfrac{1}{2} |\vec{AB} \times \vec{AC}|$$

parallelogram law of addition of vectors

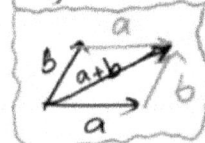

**Scalar product**

$$U = \begin{pmatrix} U_1 \\ U_2 \\ U_3 \end{pmatrix} \qquad V = \begin{pmatrix} V_1 \\ V_2 \\ V_3 \end{pmatrix}$$

$$U \cdot V = \begin{pmatrix} U_1 V_1 \\ U_2 V_2 \\ U_3 V_3 \end{pmatrix}$$

$$U \cdot V = |U| \times |V| \times \cos\theta$$

$$\therefore \cos\theta = \frac{U \cdot V}{|U| \times |V|}$$

**Vector equation of line**

$$r = a + tb$$

Position vector / direction vector

$$U = 2\hat{i} + 4\hat{j} - 5\hat{k}$$

$$V = 3\hat{i} - 7\hat{j} - 8\hat{k}$$

$$U \cdot V = \begin{bmatrix} 2 \\ 4 \\ -5 \end{bmatrix} \cdot \begin{bmatrix} 3 \\ -7 \\ -8 \end{bmatrix} = 6 - 28 + 40 = 18$$

If $U \cdot V > 0$ acute angle

$U \cdot V = 0$ Right angle

$U \cdot V < 0$ Obtuse angle

Equation of Plane using Normal Vector

$$r \cdot n = a \cdot n$$

Cartesian Equation of a plane

$$ax + by + cz = d$$

$$P(Bi/A) = \frac{P(Bi) \cdot P(A/Bi)}{P(B_1)P(A/B_1) + P(B_2)P(A/B_2) + P(B_3)P(A/B_3)}$$

THREE EVENTS

BAY'S THEOREM

TWO EVENTS

$$P(B/A) = \frac{P(B) \cdot P(A/B)}{P(B)P(A/B) + P(B')P(A/B')}$$

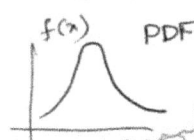 $f(x)$ PDF

$$\int_{-\infty}^{\infty} f(x)dx = 1$$

$$\mu = E(X) = \int_{-\infty}^{\infty} x \, f(x) dx$$

PROBABILITY DISTRIBUTION FUNCTION

MODE
Max of $f(x)$ is mode

Median = m

$$\int_{-\infty}^{\infty} f(x)dx = \frac{1}{2}$$

$$Var(X) = E(x^2) - \{E(X)\}^2$$

$$= \int_{-\infty}^{\infty} x^2 f(x) dx - \mu^2$$

## SOME MORE DERIVATIVES

$$\frac{d}{dx}\tan x = \sec^2 x$$

$$\frac{d}{dx}(\sec x) = \sec x \tan x$$

$$\frac{d}{dx}(\csc x) = -\csc x \cot x$$

$$\frac{d}{dx}(\cot x) = -\csc^2 x$$

$$\frac{d}{dx}(a^x) = a^x \cdot \ln a$$

$$\frac{d}{dx}\log_a x = \frac{1}{x \cdot \ln a}$$

$$\frac{d}{dx}\sin^{-1}x = \frac{1}{\sqrt{1-x^2}}$$

$$\frac{d}{dx}\cos^{-1}(x) = -\frac{1}{\sqrt{1-x^2}}$$

$$\frac{d}{dx}\tan^{-1}x = \frac{1}{1+x^2}$$

## SOME MORE INTEGRALS

$$\int a^x\, dx = \frac{1}{\ln a}\cdot a^x + C$$

$$\int \frac{1}{a^2+x^2}\, dx = \frac{1}{a}\tan^{-1}\left(\frac{x}{a}\right) + C$$

$$\int \frac{1}{\sqrt{a^2-x^2}}\, dx = \sin^{-1}\left(\frac{x}{a}\right) + C$$

### Integration by Parts

$$\int u\frac{dv}{dx}\, dx = uv - \int v\frac{du}{dx}\cdot dx = \int u\, dv = uv - \int v\, du$$

$$I = \int \sin x \cdot e^x\, dx \quad (\text{Repeated Integrals})$$
$$= \sin x \cdot e^x - \int \cos x \cdot e^x\, dx$$
$$= \sin x \cdot e^x - \left[\cos x \cdot e^x - \int -\sin x\, e^x\, dx\right]$$
$$I = \sin x \cdot e^x - e^x \cos x - I \Rightarrow I = \frac{e^x}{2}(\sin x - \cos x)$$

eg $\int \ln x \cdot 1$

$= \ln x \cdot x - \int \frac{1}{x}\cdot x\, dx$

$= x \cdot \ln x - x + C$

## Differential Equation

**EULER'S METHOD** $\rightarrow y_{n+1} = y_n + hf(x_n, y_n)$ where $x_{n+1} = x_n + h$

### Variable Separable

$$\frac{dy}{dx} = 2x$$

$$\int dy = 2\int x\, dx$$

$$y = 2\frac{x^2}{2} + C$$

$$y = x^2 + C$$

### Maclaurin Series

$$f(x) = f(0) + x f'(0) + \frac{x^2}{2!}f''(0) + \cdots$$

$$e^x = 1 + x + \frac{x^2}{2!} + \cdots$$

### Homogeneous form

$$\frac{dy}{dx} = f(y/x)$$

Sub $y = vx \Rightarrow dy/dx = x\frac{dv}{dx} + v$

$$\frac{dy}{dx} = \frac{x+2y}{x} = 1 + 2\left(\frac{y}{x}\right)$$

$$\therefore x\frac{dv}{dx} + v = 1 + 2\left(\frac{vx}{x}\right)$$

$$x\frac{dv}{dx} = 1 + v$$

$$\int \frac{1}{1+v}\, dv = \int \frac{1}{x}\, dx$$

$$\ln|1+v| = \ln|x| + C$$

$$y = Ax^2 - x$$

### Integrating factor

$$\frac{dy}{dx} + P(x)y = Q(x)$$

Multiply

$$I(x) = e^{\int P(x)\, dx}$$

$$\frac{dy}{dx} + 2y = e^x$$

$$e^{2x}\frac{dy}{dx} + 2e^{2x}y = e^{3x}$$

$$\int \frac{d}{dx}\, y \cdot e^{2x} = \int e^{3x}$$

$$y\, e^{2x} = \frac{1}{3}e^{3x} + C$$

$$y = \frac{e^x}{3} + \frac{C}{e^{2x}}$$

**Understanding the IA Criterions and thoroughly understanding their ingredients.**

Here following points should be kept scoring optimal in each criterion. The examples for each criterion have been mentioned for each criterion.

**Presentation criterion A:**

0 The exploration does not reach the standard described by the descriptors below.

1 The exploration has some coherence or some organization.

2 The exploration has some coherence and shows some organization.

3 The exploration is coherence and well organized.

4 The exploration is coherent, well organized, and concise.

Presentation is one of the most essential components of the IA as it overall develops an impression on the examiner. To score optimal in this criterion following things should be kept in mind as this what an examiner mainly looks for.

✓ Front page should have the title and number of pages. Here are the few examples.

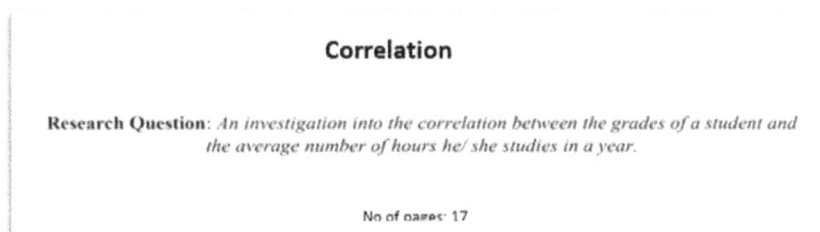

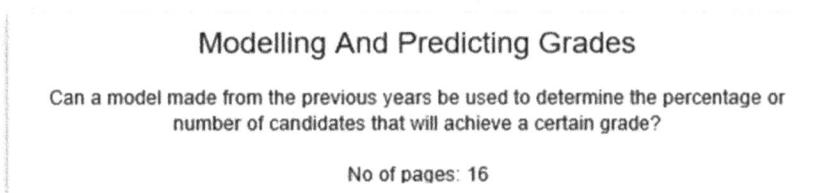

**Correlation**

**Research Question**: *An investigation into the correlation between the grades of a student and the average number of hours he/ she studies in a year.*

No of pages: 17

**Modelling And Predicting Grades**

Can a model made from the previous years be used to determine the percentage or number of candidates that will achieve a certain grade?

No of pages: 16

✓ 12-20 pages with double line spacing
✓ Page numbering
✓ Introduction
✓ Specific Aim: In general, presentation means the overall organization and coherence of the IA. If your classmate can understand and follow your IA the way you understand it, it means, your IA is well presented. Being an examiner, we look for three things in the introduction, i.e., aim, rationale and plan. Given below are the two examples for the clear aim and an unclear aim in the IA.

✓ Focused Rationale

✓ Transparent Plan. There should be a clear approach and plan given in the introduction as mentioned below

**Plan and Approach**

To answer this question only secondary data could be used which was provided in the IB statistical bulletins that are present for anyone to view.

My first instinct was to take the data from the last 5 years and take the averages of the data after which I would plot it on a cumulative frequency curve but there were a few problems with that. Firstly, the May and November examinations had slightly different pass rates and mean grades which made it so that an average of both the sessions wouldn't be able to capture the most accurate picture of the last five years.

The second problem I found was that I needed to choose whether to plot the kids who didn't receive the diploma or the ones that did because in the statistical bulletins the percentages of each are different. Since all my seniors obtained the Diploma, I decided that the latter was a better option.

Listening to their grades the expected cumulative frequency curve would follow a logarithmic curve of $y = a\ln x + b$ with a vertical stretch factor of a and a y-intercept of b.

To test my hypothesis, I decided to take a larger data set and then use that to see if it would follow the proposed path as more people are near the middle leading to a steeper gradient at the start which slowly reduces to become an asymptote at 100% in an ideal case.

✓ The concepts should be elaborated clearly (avoid leaving the reader in the grey area)
✓ The transition between the topics and distinct concept should be linked (Fluidity of the document should be clear). Before starting a new paragraph, make your reader prepared what is going to come in the next paragraph and how it is related to the previous one by providing a rationale behind it.
✓ Avoid adding long table (Prefer appendices)
✓ Avoid repetition: This happens mostly in the IA and students do not take this into account and lose score in lacking conciseness of the IA.

✓ Conclusion in line with aim

## Conclusion

It can be deduced that the data does follow a normal distribution however the scores of the students that have obtained the Diploma lie well in between three standard deviations of the mean and so the third empirical rule doesn't hold a lot of value in this case. Any how it came as a surprise that the point distribution does follow somewhat of a normal distribution and thus using the normal CDF function I will try and answer my question of: was the grade distribution for the May 2020 exams like previous years? And how closely does it resemble it.

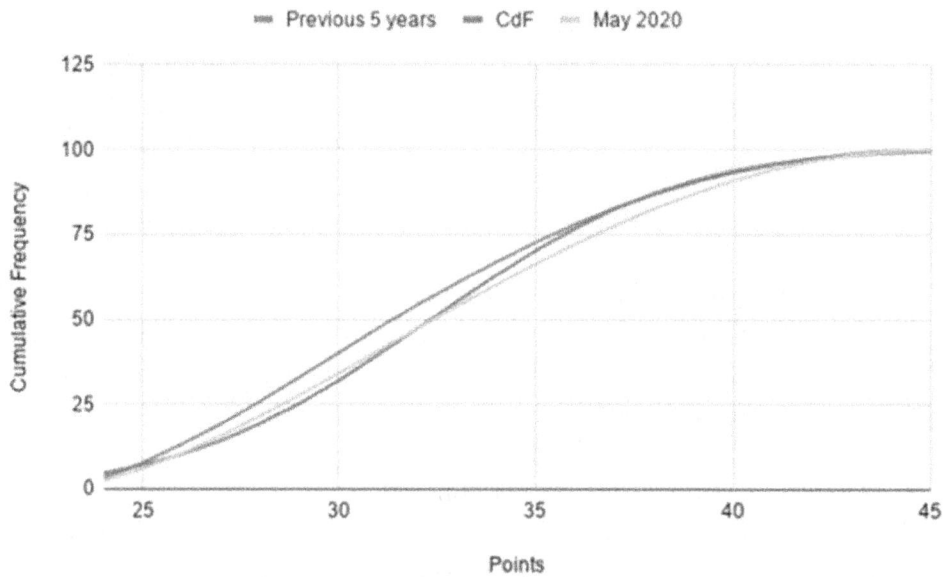

Fig. 10 CDF compared to the May 2020 and Previous 5 May examinations

This is a very interesting site to see. The CDF line follows the curve of the May 2020 examinations at the beginning and then merges with the line of the previous 5 May examinations. After seeing this I am confident that the grade distributions follow a normal distribution to a large extent and no specific points were awarded out of proportion to the May 2020 batch. Using excel I found the coefficient of determination $r^2$ of the Cdf with the previous 5 years to be 0.993465089 while for the May 2020 examinations it turned out to be 0.99616521. Although a minute difference it is safe to say that the CDF model is very precise at predicting the number of candidates receiving a particular grade.

## Mathematical Communication B:

**0** The exploration does not reach the standard described by the descriptors below.

**1** The exploration contains some relevant mathematical communication which is partially appropriate.

**2** The exploration contains some relevant mathematical communication.

**3** The mathematical communication is relevant, appropriate, and mostly consistent.

**4** The mathematical communication is relevant, appropriate, and consistent throughout.

Mathematical communication is one of the most technical aspects of the IA as the examiner takes it quite seriously. By following the checklist below once can ensure maximum marks in this criterion.

- ✓ Define all key terms and variables
- ✓ Use real arithmetic sign (avoid generic signs such as * for multiplication or / for divisor)
- ✓ Write equations through an equation editor (Don't use alphabets such as use $2x + y$ instead of 2x+y)

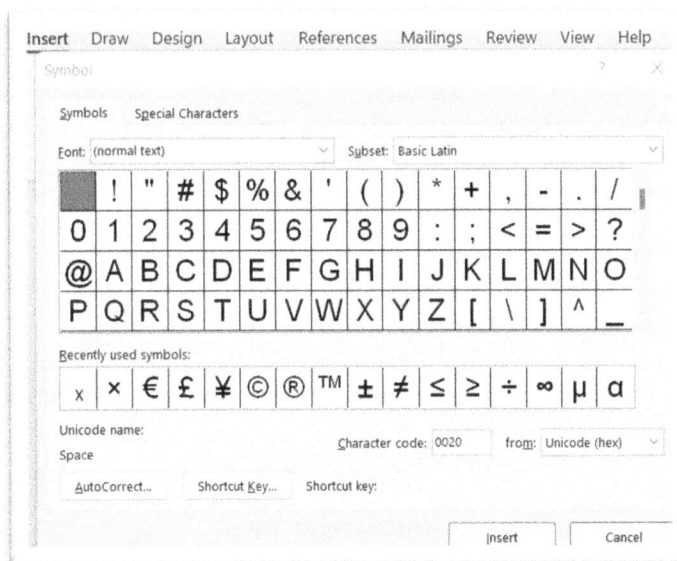

- ✓ Center align the equations
- ✓ Along with equations the chart, table, graphs etc. should also be aligned in center
- ✓ Axes should be clearly labelled along with clear title of each chart

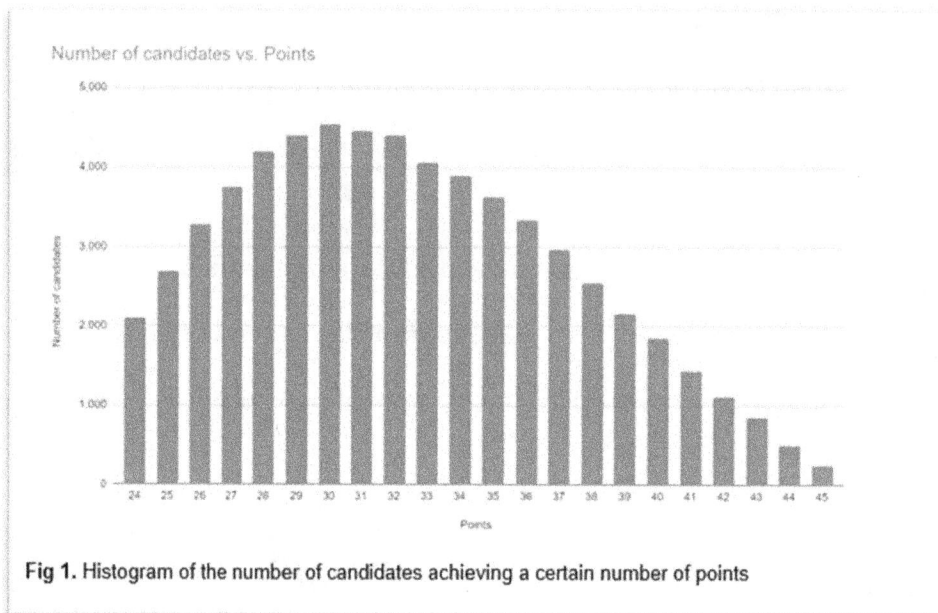

Fig 1. Histogram of the number of candidates achieving a certain number of points

✓ Keep in mind the significant figures as they should remain the same throughout the IA to maintain flow. The estimation should be quoted and done on the required spaces.

> Since I have collected the data from IB website which contains three significant figures, therefore, I will be keeping all my final answers to three significant figures. This would also allow me to keep with the practice for my final exams where I must leave my final answers to three significant figures.

> Since I have used Vernier Calipers to take my measurements, therefore I will keep all my findings and numbers to hundredth part of the mm as this is the least count of the device, I used to collect my data.

## Personal Engagement C:

**0 The exploration does not reach the standard described by the descriptors below.**

**1 There is evidence of some personal engagement.**

**2 There is evidence of significant personal engagement.**

**3 There is evidence of outstanding personal engagement.**

Personal engagement is one of the trickiest criterions as most of the students struggle to execute this criterion on merit. These are some points that can help the student attain maximum score in criterion C.

✓ Affiliation with the topic and expectation of the results should be evident throughout the document
✓ Shouldn't be very generic
✓ An anecdote should complement the topic selection.

34

✓ Collection of data could be secondary or could be primary however collection method should be clear and straightforward.

> **Data Collection and Results**
>
> Using the statistical bulletins provided by the IB, I took the data from the last 5 May examination series from 2015-2019. The reason being that taking into account November series leads to data being skewed since the change in number of candidates is drastic which gives slightly different results. I have taken the averages of the number of candidates and then calculated the percentage and the cumulative frequency.
> The boundaries of each data point are the number of points the students have achieved while obtaining a diploma, i.e. 24 to 45.

✓ Opinion on the topic should be given at times as the flow of the document shouldn't seem mechanical. You do have the write to wonder or agree to disagree at times. (The document should seem realistic as in everything can't be perfect)
✓ Open-mindedness should surface the document (Perspective's should be welcomed)
✓ The topic shouldn't be very generic (Prefer something unique)
✓ Consistently reinforce why are you glad and excited to work on a certain topic.

## Reflection D:

**0 The exploration does not reach the standard described by the descriptors below.**

**1 There is evidence of limited reflection.**

**2 There is evidence of meaningful reflection.**

**3 There is substantial evidence of critical reflection.**

Reflection is one of the most meticulous criterions where students are required to go deep and evaluate the process on merit. Keeping in mind the following points can bear maximum points in this criterion.

✓ Reflection must occur consistently at each point as it is one of the most common mistakes most student make, as they just restrict it to the evaluation part

✓ Comment on whatever the findings are
✓ Contextualize each result (as in what does r value imply, what does an intercept indicate etc.)
✓ Be honest and discuss the limitations of the process before examiner identifies it
✓ Evaluate the process in term of difficulty and how the experience was.
✓ Mention strengths and weaknesses
✓ Add various approaches and possible extensions.

### Use of Mathematics E:

0 The exploration does not reach the standard described by the descriptors below.

1 Some relevant mathematics is used.

2 Some relevant mathematics is used. Limited understanding is demonstrated.

3 Relevant mathematics commensurate with the level of the course is used. Limited understanding is demonstrated

4 Relevant mathematics commensurate with the level of the course is used. The mathematics explored is partially correct. Some knowledge and understanding are demonstrated.

5 Relevant mathematics commensurate with the level of the course is used. The mathematics explored is mostly correct. Good knowledge and understanding are demonstrated.

6 Relevant mathematics commensurate with the level of the course is used. The mathematics explored is correct. Thorough knowledge and understanding are demonstrated.

Criterion E is a make or break for an IA as this determines the caliber and the rigor that has been depicted throughout the process. Following pint can prove to be a grade booster for the IA.

- ✓ Relevance is the key (Randomly discussing a certain concept is detrimental to the fluidity of the document)
- ✓ Clear knowledge and understanding of the applied concept should be demonstrated.
- ✓ The level should be kept on view as in an HL student cannot use a simple mathematical process such as 'finding correlation'. At least two to three concepts should be demonstrated
- ✓ Prefer showing working for each step
- ✓ Be precise and logical
- ✓ Avoid using simple math or just prior knowledge.
- ✓ Clear arguments should be made.

## Radial Cycle for the Criterion

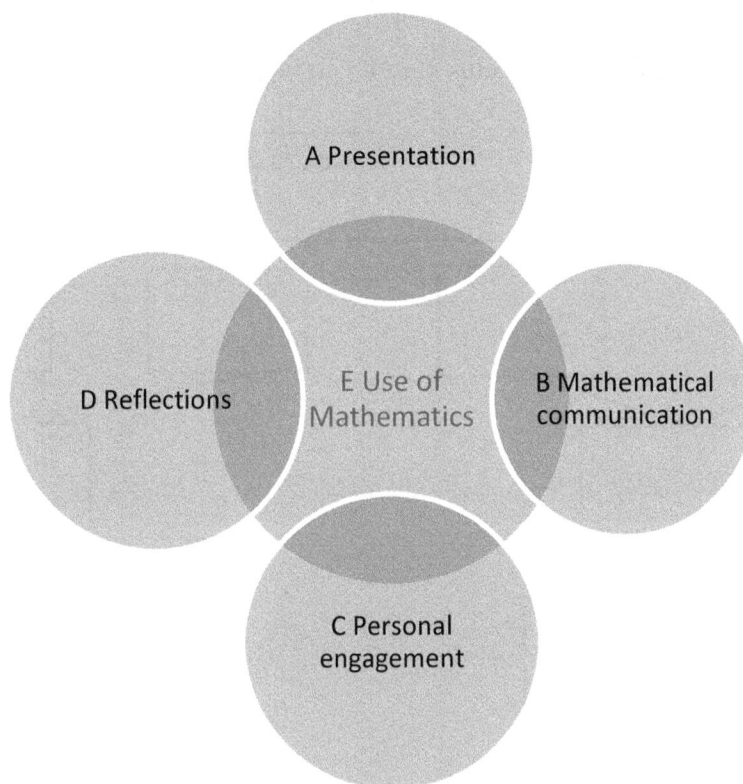

When I was new to IBDP, I used to think that one can get high score in the IA if one scores high in the first four criterions which are general and contain 14 scores out of 20 in total. And then getting 3 out of 6 in criterion A will allow them to score low high. However, later I discovered that criterion E is the core of all the other criterions. I mean how can one score high in presentation, communication, personal engagement, and reflections if the math applied in executing the exploration does not have rigor in it, or it may not be sufficient to meet the level of criterion D. If we do not use rigorous math, we won't get enough of the opportunities to make use of symbols, equations, and mathematical jargons. We can't go deeper into the exploration. We have not much to reflect on to get high scores in the first four criterions. So, while the process of selection of topic, make sure the topic has enough of the depth in it which would allow us to apply appropriate and rigorous math procedure in it to get nothing less than 5 in the use of math, only then we would be able to score good in the other four criterions.

Generally, math IAs can be done in the following areas but not limited to this.

## Mathematical modelling

Mathematical modelling is the vast area and it usually considered as the high scoring IAs as it may incorporates all the five topics (number and algebra, functions, geometry and trigonometry, probability and statistics and calculus in it. It can be further divided as.

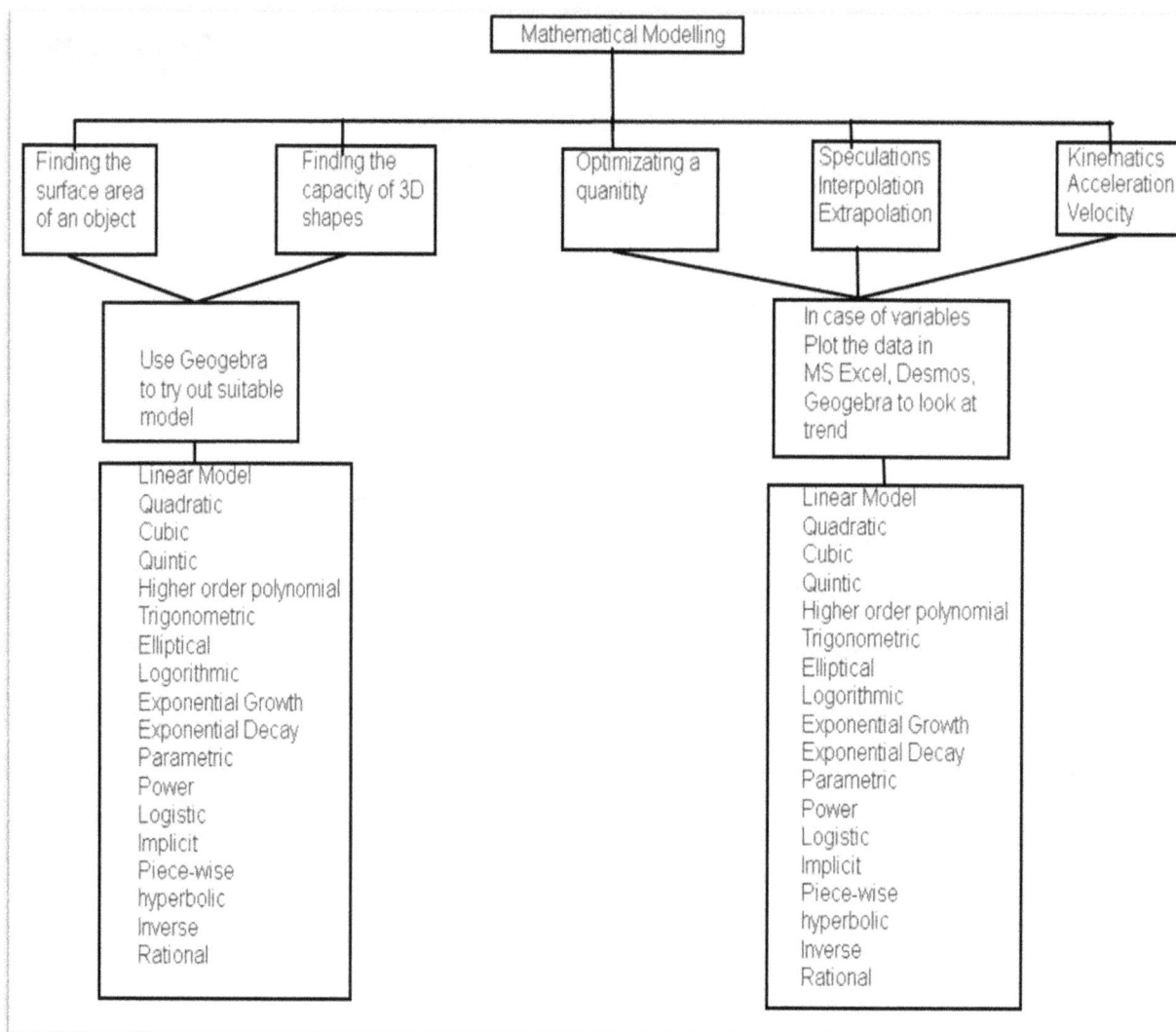

```
                          ┌─────────────────────────┐
                          │  Mathematical Modelling │
                          └─────────────────────────┘
```

| Finding the surface area of an object | Finding the capacity of 3D shapes | Optimizating a quanitity | Speculations Interpolation Extrapolation | Kinematics Acceleration Velocity |

**Use Geogebra to try out suitable model**

**In case of variables Plot the data in MS Excel, Desmos, Geogebra to look at trend**

| | |
|---|---|
| Linear Model | Linear Model |
| Quadratic | Quadratic |
| Cubic | Cubic |
| Quintic | Quintic |
| Higher order polynomial | Higher order polynomial |
| Trigonometric | Trigonometric |
| Elliptical | Elliptical |
| Logorithmic | Logorithmic |
| Exponential Growth | Exponential Growth |
| Exponential Decay | Exponential Decay |
| Parametric | Parametric |
| Power | Power |
| Logistic | Logistic |
| Implicit | Implicit |
| Piece-wise | Piece-wise |
| hyperbolic | hyperbolic |
| Inverse | Inverse |
| Rational | Rational |

## Finding the Surface Area of a Building

A student may think of modelling someone's face, picture, building using GeoGebra, Desmos or any suitable tool and find the best model that fits the situation. AA students tends to use visuals and derive models using standard forms and show all the working stepwise to demonstrate their understanding and use of math. However, AI students can also derive models however, their focus should be on interpretation and analysis of that model. Here is an example of GEMS WORLD ACADEMY where the aim of the exploration is finding the surface area of the front of the building using GeoGebra and then using integration to workout area under the curve.

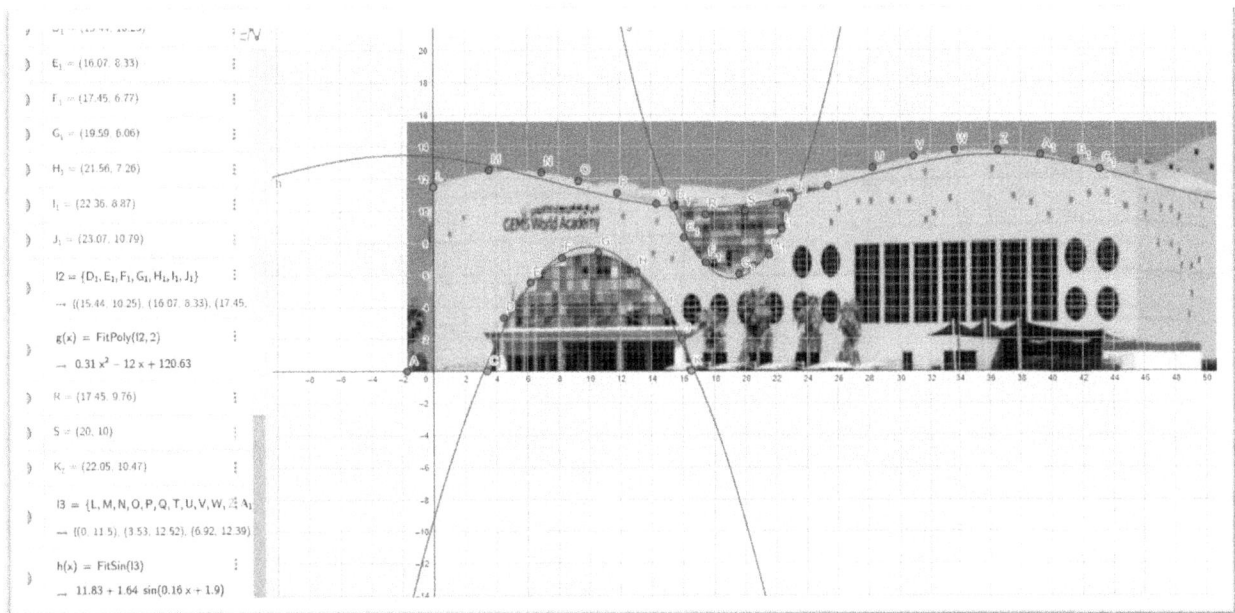

Area under and above the two quadratics can be found using https://www.symbolab.com/. Areas of circular and rectangular windows can be worked out using their formulas. Subtracting this area from sum of all these areas under the sine model will be the required surface area of the front of the building.

Area under the curve for parabola facing down

$$\int_{15}^{23.7}\left(0.31x^2 - 12x + 120.63\right)dx = 56.16981$$

Area under the curve for parabola facing upward

$$\int_{3.2}^{16.8}\left(-0.17x^2 + 3.39x - 9.1\right)dx = 70.4443722$$

Area under the sine curve

$$\int_{0}^{44}\left(1.64\sin\left(0.16x + 1.9\right) + 11.83\right)dx = 526.27526\ldots$$

**Find the Volume of Revolution of 3D Shapes**

Finding the volume of 3D objects can also be done using GeoGebra. Here is an example of finding the volume of a vase. The first step is to model the object. Try out best model that fits the situation. I tried sinusoidal first to see if this fits the best. However, you can see the model looks a little off between points D to G and then from P to Q. We can try another.

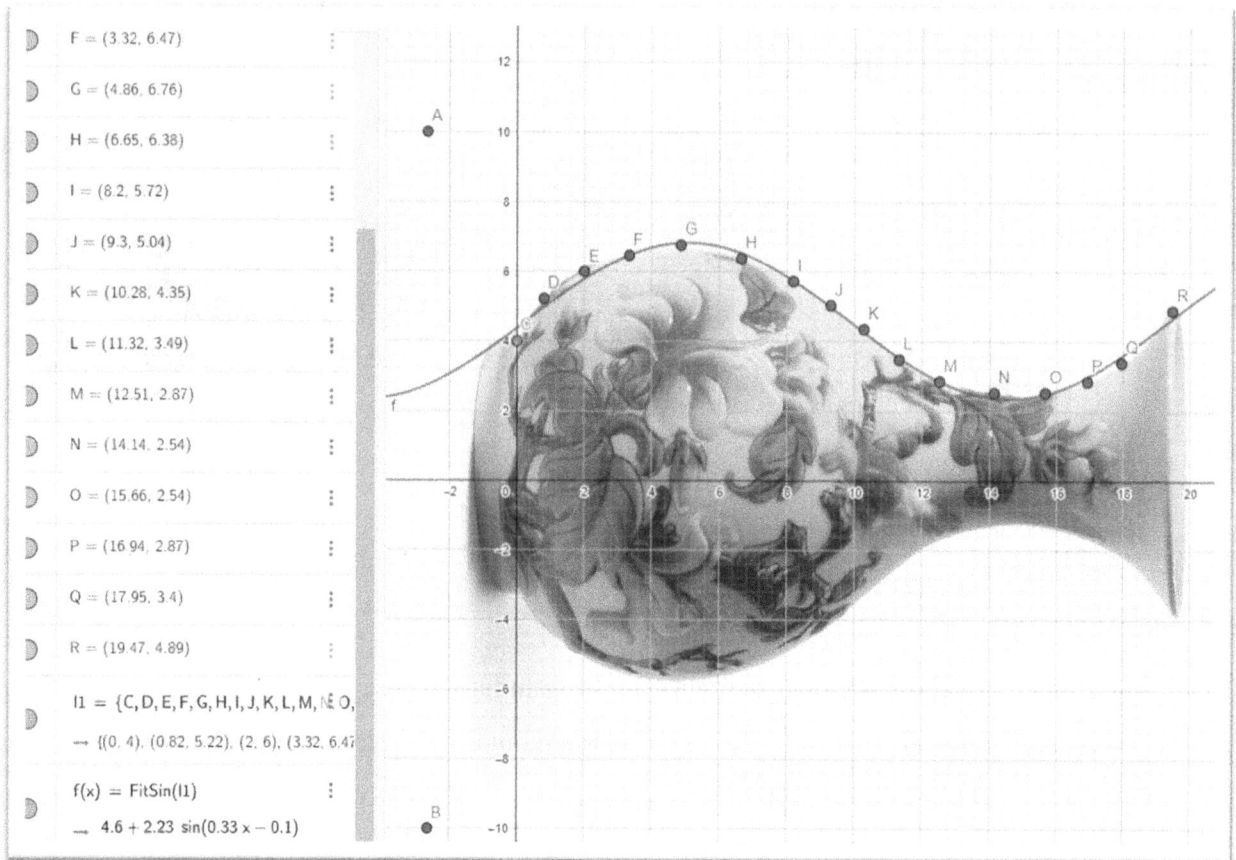

F = (3.32, 6.47)

G = (4.86, 6.76)

H = (6.65, 6.38)

I = (8.2, 5.72)

J = (9.3, 5.04)

K = (10.28, 4.35)

L = (11.32, 3.49)

M = (12.51, 2.87)

N = (14.14, 2.54)

O = (15.66, 2.54)

P = (16.94, 2.87)

Q = (17.95, 3.4)

R = (19.47, 4.89)

l1 = {C, D, E, F, G, H, I, J, K, L, M, N, O,

→ {(0, 4), (0.82, 5.22), (2, 6), (3.32, 6.47)

f(x) = FitSin(l1)

→ 4.6 + 2.23 sin(0.33 x − 0.1)

Sometimes the model looks very off, and we can reject that model without applying any statistical tool. However, sometimes, the model looks very close in which case we may look for coefficient of determination. The coefficient of determination is **a measurement used to explain how much variability of one factor can be caused by its relationship to another related factor.** This correlation, known as the "goodness of fit," is represented as a value between 0.0 and 1.0. The closer the value is to 1, the better the model is and vice versa. If we look at the trend, cubic function can be a good fit here.

Given below is the cubic model which looks better than sine model visually, however, we will find the coefficient of determination to see which model is best for this vase.

40

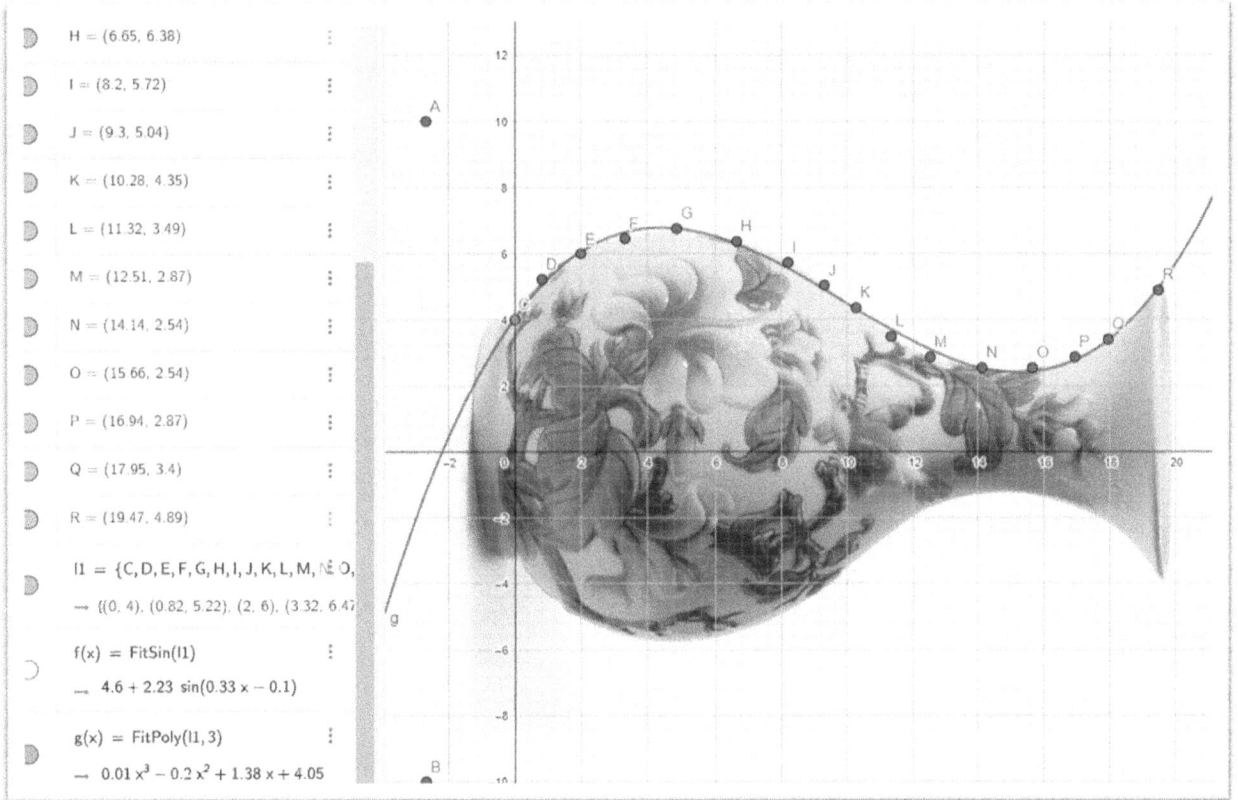

Volume of revolution can be found by putting functions in https://www.symbolab.com/. Since we found that cubic function is best to

$$\pi \cdot \int_0^{44} \left(0.01x^3 - 0.2x^2 + 1.38x + 4.05\right)^2 dx = 4811397.83624\ldots$$ model

this, this is how we can find the volume of this vase.

Volume of vase =

Symbolab allows us to show the whole process step-by-step which is important for AA students to show and demonstrate how to get to final answer. Below is the 3D image of vase that we wanted to calculate volume for,

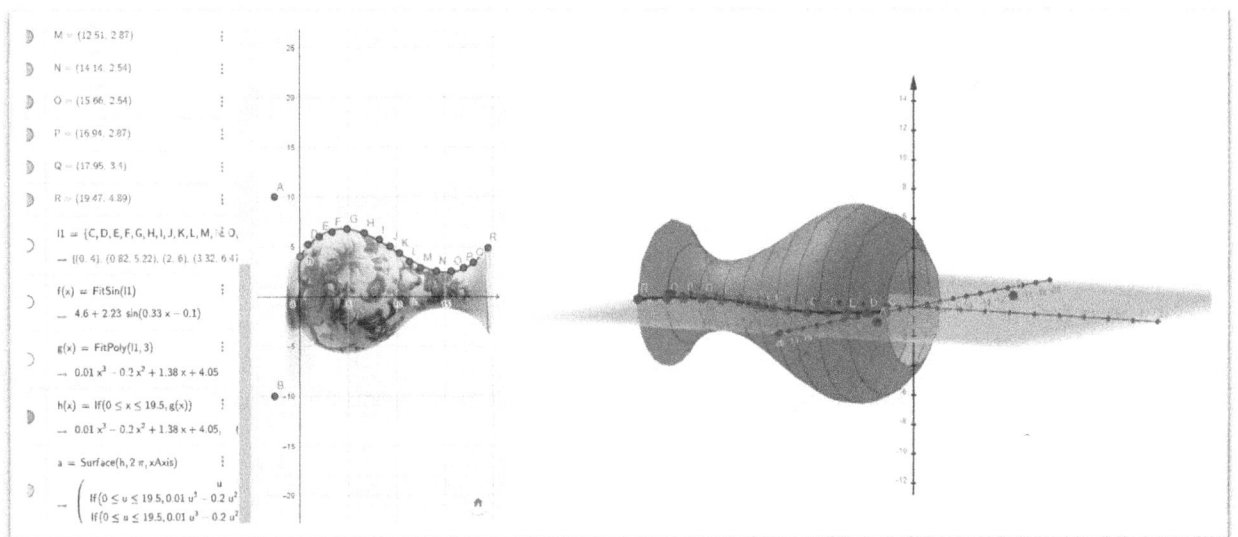

J = (-1.35, 7.63)

K = Point(yAxis)

→ (0, 7.74)

L = (1.71, 7.68)

M = (3.26, 7.36)

N = (5.1, 6.58)

O = (6.47, 5.54)

P = (7.66, 4.12)

Q = (8.37, 2.45)

R = (8.58, 0.13)

l1 = {C, D, E, F, G, H, I, J, K, L, M, N, O,

→ {(-8.31, 0.19), (-8, 2), (-7.63, 3.82), (-

$f(x) = \text{FitPoly}(l1, 2)$

→ $-0.09 x^2 + 0.02 x + 8.41$

$g : y = 7.74 \sqrt{1 - \dfrac{x^2}{8.58^2}}$

## Correlation

Correlation IAs are generally recommended for SL students however, if we use non-linear/transcendental functions in modelling the two variables, this may work for HL too. In order to look for the linear correlation between the two variables, we use Pearson's Product Moment correlation commonly known as 'r' value ranging from -1 to +1.

**Facts about $r$**

- The value of $r$ is always in the interval $-1 \leqslant r \leqslant +1$
- The sign of $r$ tells us the direction of the correlation: positive or negative or zero.
- The size of $r$ tells us the strength of the linear correlation, as shown.

| Perfect negative correlation | No correlation | Perfect positive correlation |
|---|---|---|
| $r = -1$ | $r = 0$ | $r = +1$ |

$$-1 \qquad\qquad 0 \qquad\qquad 1$$

| Strong negative correlation | Weak negative correlation | Weak positive correlation | Strong positive correlation |

- If $r = +1$, there is a perfect positive linear correlation; all the points fall on a line with positive slope.
- If $r = 0$, there is zero correlation.
- If $r = -1$, there is a perfect negative linear correlation; all the points fall on a line with negative slope.
- $r$ has no units, and is not a percentage.

If the correlation is not linear, we use Spearman's rank correlation coefficient, $r_s$,

**Facts about $r_s$**

- The value of $r_s$ is always in the interval $-1 \leqslant r \leqslant +1$
- The sign of $r_s$ tells us the direction of the rank-order correlation: positive or negative or zero.
- The size of $r_s$ tells us the strength of the rank-order correlation.
- If $r_s = +1$, there is a perfect positive rank-order correlation.
- If $r = 0$, there is zero correlation.
- If $r = -1$, there is a perfect negative rank-order correlation.
- $r$ has no units, and is not a percentage.
- $r$ is a measure of monotonicity.

## Data Analysis

### Voronoi Diagrams

Voronoi diagrams can be used to find the best location to open a store. For example, if we want to open a coffee shop in London, we can spot all the coffee shops in vicinity and can work out the place that is at a maximum possible distance from all the existing coffee shops in the town.

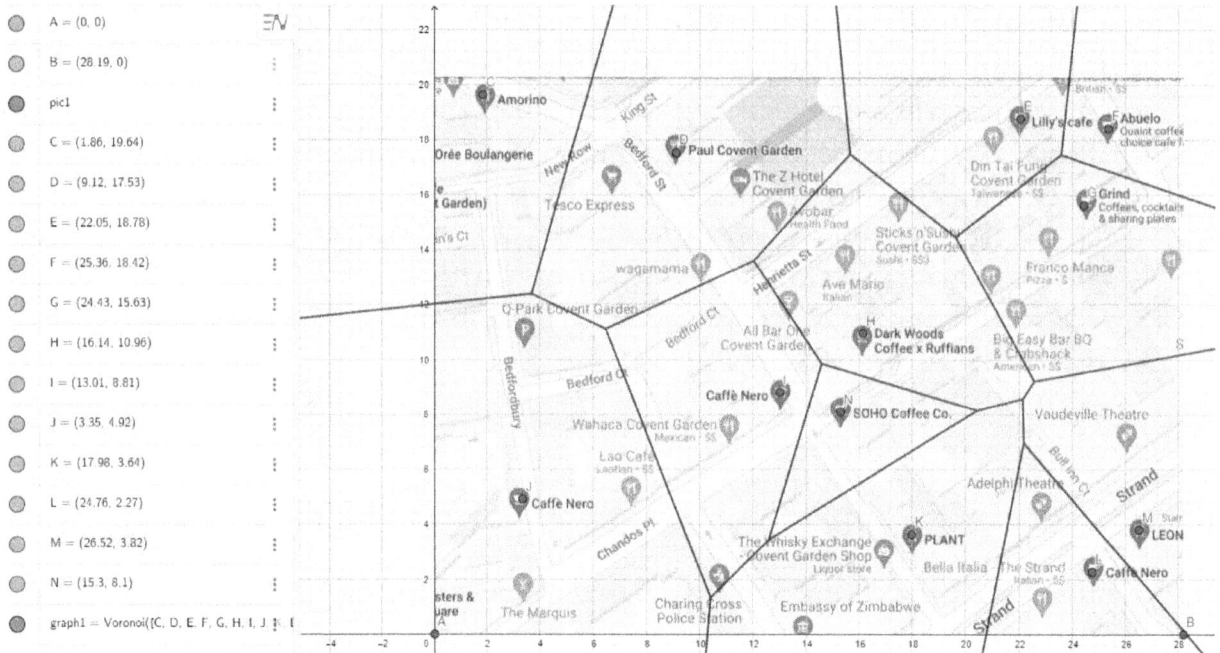

A = (0, 0)
B = (28.19, 0)
pic1
C = (1.86, 19.64)
D = (9.12, 17.53)
E = (22.05, 18.78)
F = (25.36, 18.42)
G = (24.43, 15.63)
H = (16.14, 10.96)
I = (13.01, 8.81)
J = (3.35, 4.92)
K = (17.98, 3.64)
L = (24.76, 2.27)
M = (26.52, 3.82)
N = (15.3, 8.1)
graph1 = Voronoi({C, D, E, F, G, H, I, J, K, L})

## Probability and Distributions

In statistics, IAs can be picked from normal distribution, binomial distribution, poison distribution and probability density functions.

Phenomenon which are normally distributed are

- Height
- Rolling some dice
- Tossing a coin
- IQ level of students
- SAT score
- IB Score
- Stock Market
- Income distribution in economy
- Size of the shoes
- Birth Weight
- Weight of canned juice
- Bag of cookies

And many more phenomenon which can lead to normal distribution. Phenomenon which can be modelled using binomial distribution are as follows

- Number of people responding 'Yes' to a particular survey.
- Number of patients responding to a specific medicine
- Flipping a coin for n times where n is greater or 50.
- Number of games win by a team in a tournament.
- Number of votes won by a candidate in an election.

44

- Number of defective products in a production run.

An exemplar.

My first objective is to produce a probability distribution function, to determine the probability of a certain coach having a certain number of years. I started off by producing a table using the data collected above. The table shows the number of coaches in my sample that were head coach within a certain interval of time (in years). For example, the number of coaches from the data collected above that spent less than a year coaching was 17. The length of each interval was kept equal, and the length of

Using the obtained probability for each interval, we can produce a graph representing the plots of the probability distribution function that we will define later on in this exploration:

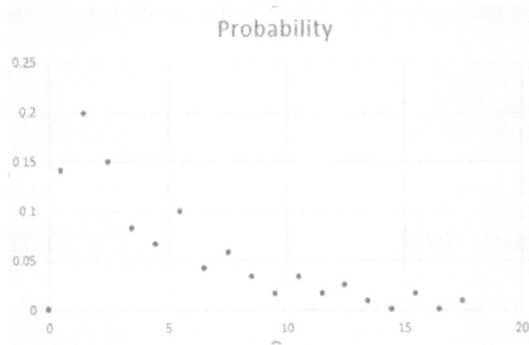

Probability

Two different functions were made through the use of technology (excel), and a PDF was produced as a result. The PDF is:

$$f(x) = \begin{cases} 0 & x <= 0 \\ 0.0003x^5 - 0.0066x^4 + 0.0588x^3 - 0.2404x^2 + 0.3828x + 0.0006 & 0 < x <= 5.379 \\ 0.0008x^2 - 0.0238x + 0.1859 & 5.379 < x <= 18 \\ 0 & x > 18 \end{cases}$$

And the following is the graph, made on Desmos, of the PDF:

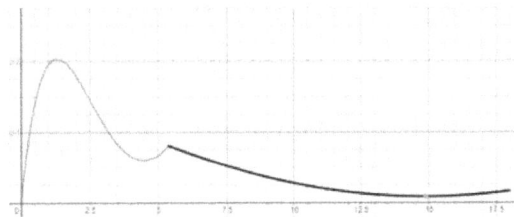

each interval was chosen by considering what number of intervals would allow us to produce a somewhat compact and still somewhat precise table. 1 year seemed to be the most reasonable interval length. The frequency of the data in this case is the number of coaches that fall under a certain time interval.

## Frequently Asked Questions and Answers.

### Selection of the topic

Q1. What is the difference between math EE and math IA?

EE and IA are almost similar in terms of what they expect from student however, EE as name suggests is an extended and much more in-depth and more complex analysis of your topic, whereas, IA is much more focused, straight forward and have limited possibilities in exploration. The difference also lies in the word limit and the criterions as well.

Q2. Does the selection of topic vary for AA/AI math?

There is no such difference in terms of the selection of the topic for the math AA and AI. The difference lies in the approach and execution of the exploration. AI is more technology driven and focuses more on the application and interpretation of mathematical techniques. Whereas math AA is more of demonstrating on how formulas, equations and models are derived followed by the interpretation in the specific context. AA requires students to show all the mathematical steps involved in the derivation of their mathematical procedures.

Q3. At what level the selection of topic varies in terms of writing an IA for SL and HL?

While picking up a topic for the HL, one should make sure that the topic provides enough of the depth which allows one to apply rigorous (At least one math procedure from HL). We can apply math on almost everything but the question one should ask oneself is whether the topic meets the rigor of criterion E. In case of SL, make sure the topic should not limit to prior knowledge math.

Q4. Can two students use same topic for exploration in one class?

Yes, they can, however, the exploration must be different in terms of data, approach, and findings. They can work on two different avenues within the same topic.

### Presentation

Q1. What should be on the first page? Is table of content an important component for the IA?

First page should have a topic and a specific title of that topic, for example, putting just 'Modeling' as a topic is not sufficient. It is better to write' Can a model made from the previous years be used to determine the percentage or number of candidates that will achieve a certain grade? Table of content is not mandatory. Put no of pages.

Q2. Can we submit handwritten IA?

Students can write text, equations, and draw graphs and table by hand, however, word processed is encouraged and looks more professional. And there are lot of tools available now-a-days which help draw graphs, use symbols at ease.

Q3. Can student use external resources such as MS office, Desmos, GeoGebra etc.?

Yes of course, students can use external sources and should acknowledge these in their IA as the skills they learnt while writing the IA.

Q4. Do we need any cover page as we do in EE?

No page cover is required in the IA.

Q5. Does an exploration have to be less than 20 pages long to be concise?

Not necessarily, it depends on the topic of the IA and the approach. If justified, relevant and necessary math procedures, calculations, graphs, or interpretation are used in the IA, no of pages can go little high.

Q6. Is bibliography or appendices included in the page count?

No, it does not.


## Mathematical Communication

Q1. What should be the target audience for the student?

The target audience for this exploration should be fellow students. Student should request their fellow members to read their IAs followed by a small viva to double check if they understand the IA in similar context.

Q2. What do we mean by defining the key terms in math IA?

It means defining every term that is not commonly known, for example, define what is income per capita, GDP, scoring an Ace in tennis, body mass index etc. Don't add anything unexplained that a student must google it to know.

Q3. If the IA does not contain graph, charts, or tables, can we still earn some score in criterion B?

It is recommended to use multiple forms of mathematical procedures but not necessary to use all forms. Wherever possible, use all appropriate ways of showing information, models, equations, and formulas.

## Personal Engagement

Q1. Does personal engagement mean writing a good and convincing start why to pick a specific topic?

There is common misconception that personal engagement means student need to have a personal vested interest in the topic which leads them writing deliberate lies which can reveal on examiner at ease. Personal engagement demands the topic to be unique and independent. It should have unique approach of exploring the topic. It should display some degree of creativity by exploring the varied aspects of the IA and discuss unique avenues.

## Reflections

Q1. What constitute reflections in the IA?

Discussing the challenges faced by the students and how those overcome using various skills mentioned in the learner profile. Reflections can be done at various level:

- Reflection on how to collect the data, and if there is any specific sampling technique used and why.
- Doing commentary on your results and equations
- Discussing the limitations and assumptions, interpolation, and extrapolations.
- What went good and what bad, critically assessing it.
- Writing possible extensions of the IA.

Q2. What is the difference between a conclusion and a reflection?

A conclusion could be more descriptive and factual. It is wrapping up of what is said in the aim and rationale of the IA. Conclusion comes in the end of the IA.

Whereas reflection is an ongoing process in the exploration and must be seen throughout the work. It runs throughout the IA.

## Use of Mathematics

Q1. Should students use math that is beyond IB course?

IB does not encourage to use math beyond the IB course. If there is a need to use math outside the course, it is advisable that the level of the math used should be commensurate with the level of the IB.

Q2. What is the difference in criterion E (use of the math) in SL/HL IA?

HL students are expected to use at least one math procedure that should be purely from IB math.

Q3. How can we assure if it is a math? (Not a physics or economic IA)

Make sure you extend math in your IA. Your IA should be math heavy.

Q4. Can I derive the already derived equations like Reimann's sum?

IB expect students to apply already discovered math on real life scenarios. Proofs like Reimann's sum is already done however, if you have a different way of proving it, you can take this as your IA.

# PART II

## SEVEN EXAMPLES OF EXCELLENT INTERNAL ASSESSMENTS

The assessments featured in this section are all recently submitted IA that scored exceptionally after being moderated by the IBO. To prevent plagiarism and duplication of results, the appendices have been omitted. The IAs are presented in the exact same way as they were submitted, and without any edits or changes to formatting. We do not retain the copyright of these commentaries, nor is this publication endorsed by the IBO. The Internal Assessments are being re-printed with the permission of the original authors.

# 1. USING THE TRAPEZOIDAL RULE AND SIMPSON'S RULE TO WORK OUT THE GINI COEFFICIENT

Author: Keke Shah
Moderated Mark: 20/20
Level: Math AA HL

## Introduction

Being a student who is genuinely interested in economics, I chose it as a higher-level subject along with geography. Mathematics is a very wide discipline and contributes to many areas of studies, it also plays a crucial part in Economics. I am particularly interested in the topic of global and human development, when the topic inequality was first introduced in macroeconomics section, one of the macroeconomic objectives intrigued me: minimise wealth inequality[1].

It seized my attention because "Everyone was born unfairly.", this is what my family has invariably been talking about, this is cruel, but reality is what most of the people around the world have to face. As a result of economic growth, I believe this would widen the gap between the poor and rich, due to various reason like tax being regressive, meaning the taxes are more of a burden for the poor than the rich, which happens everywhere in our daily life when we buy products from stores and pay indirect taxes. This resulted in the poor being even poorer, the wealthy people getting even wealthier, making the poorer even harder to catch up as time passes. As I studied ahead of the macroeconomics section and bits on global development, I was introduced to the Lorenz curve, which shows the income distribution in the economy. Also, the Gini coefficient, a numerical representation of the Lorenz curve and an important tool for analysing income or wealth distribution within a country. However, it doesn't measure the actual wealth or income of the country as it only shows the relative equality in income distribution.

## Aim

As one of the government's objectives, throughout the years the problem of inequality has become more severe, therefore the Gini coefficient has also increased. I wanted to use the Lorenz curve to find the Gini coefficient of the country. In this internal assessment, I will use 2 different methods of numerical integration: Trapezoidal rule and Simpson's rule to find the Gini coefficient of the countries and evaluate the accuracy of the 2 integration methods, by comparing Gini coefficient worked out and compare it to data from THE WORLD BANK.

---

[1] https://www.economicshelp.org/blog/419/economics/conflicts-between-policy-objectives/

Background information

Lorenz curve[2]

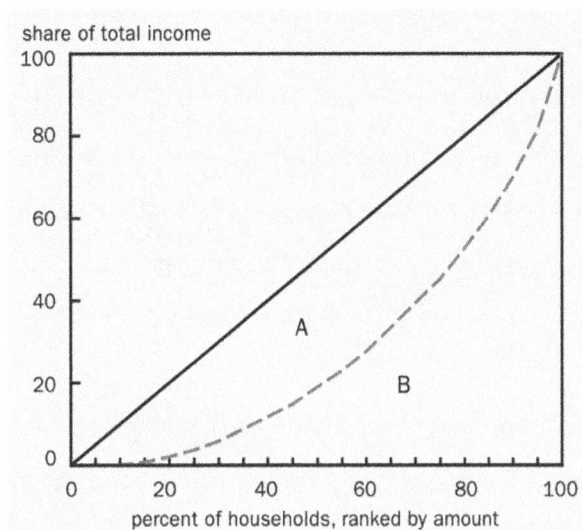

*Graph 1: an example of Lorenz curve*

Lorenz curve is a hypothetical way to depict the equality of a country by illustrating the perfect equality line as a straight diagonal line ($y = x$) shown on graph 1, and the blue dotted curved line is the Lorenz curve, which shows the actual income distribution of a country. Both axes are cumulative and measured in percentage %, with the x-axis being the percentage of households(population) and the y-axis being the shared percentage for income distribution within a country up to 100 %. To construct the Lorenz curve, we will need to first sort the households by their income, starting with the ones with the lowest income, then plot the relationship between the cumulative percentage of the population on the x-axis against the cumulative proportion of total income earned by the cumulative percentage of the population on the y-axis.

Gini coefficient

The Gini coefficient was developed upon the Max Lorenz's graph, the proportion of area under the hypothetical perfect equality line ($y = x$) and area between the 2 curves is the Gini coefficient. The value of the Gini coefficient varies between 0 to 1, 0 being the case of perfect equality, meaning that every household earns the same, and 1 in the case of perfect inequality meaning one household earns all the income in the country. On graph 1, The total

_____

[2] https://users.nber.org/~denardim/research/DeNardi_Income_Inequality_Five_Countries.pdf

area under $y = x$ is A+B, area under the blue dotted curve is B. The area between $y = x$ and Lorenz curve is A. Gini coefficient is defined as the ratio of area A to area A + B $(\frac{A}{A+B})$.

## Data collection

The sample size is all the countries in the world, since there're around 195 countries and the sample size is very large, the sampling method I will use to choose the countries to work out their Gini coefficient will be stratified sampling. I will categories the countries in terms of their geographic locations and choose 5 countries from each 6 continents (Asia and Australasia/Oceania, South America Africa, Europe, North America).

There were a lot of limitations to my data while I was collecting them. Antarctica is excluded as there're very few people living there, and Australasia/Oceania will be combined with Asia as it does not have much data updated for its countries. As the data for income distribution was very hard to find, I chose THE WORLD BANK to use, also the official released data of Gini coefficient I will use for evaluating the accuracy of integration method is from here[3]. (The Gini index from THE WORLD BANK= Gini coefficient x 100= $\frac{A}{A+B}$ on graph 1.)

However, the data for the income distributions from some counties aren't from recent years (some countries are still at 2011). Therefore, there're concerns about the bias that some countries tend to have a more reliable data. To deal with the countries with limited updated data, because this exploration aims to look at the 2 methods of numerical integration, therefore as long as the income distributions are from the same year, the values and calculations will still be valid in this case.

## Data[4]

All the 30 countries that I will use are in table 1 to 5, column B shows the year which the data will be extracted from. Column C shows the Gini coefficient from THE WORLD BANK (will be used for later comparison to work out percentage error of integrating methods).

[3] https://data.worldbank.org/indicator/SI.DST.05TH.20
[4] https://www.worldometers.info/geography/how-many-countries-are-there-in-the-world/

*Table 1: countries from Asia and Australia/ Oceania combined*

| Column A: Country | Column B: Year of data that I will be collecting from | Column C: Gini coefficient |
|---|---|---|
| 1. China | 2016 | 0.385 |
| 2. Mongolia | 2018 | 0.327 |
| 3. Japan | 2013 | 0.329 |
| 4. Indonesia | 2018 | 0.378 |
| 5. Russia | 2018 | 0.375 |
| 6. Malaysia | 2015 | 0.410 |
| 7. Turkey | 2018 | 0.419 |
| 8. Vietnam | 2018 | 0.357 |
| 9. United Arab Emirates | 2014 | 0.325 |
| 10. Australia | 2014 | 0.344 |

*Table 2: countries chosen from South America*

| Column A: Country | Column B: Year of data that I will be collecting from | Column C: Gini coefficient |
|---|---|---|
| 11. Ecuador | 2018 | 0.454 |
| 12. Honduras | 2018 | 0.521 |
| 13. Brazil | 2018 | 0.539 |
| 14. Colombia | 2018 | 0.504 |
| 15. Uruguay | 2018 | 0.397 |

*Table 3: countries chosen from Africa*

| Column A: Country | Column B: Year of data that I will be collecting from | Column C: Gini coefficient |
|---|---|---|
| 16. Botswana | 2015 | 0.533 |
| 17. South Africa | 2014 | 0.630 |
| 18. Sudan | 2014 | 0.342 |
| 19. Tanzania | 2017 | 0.405 |
| 20. Namibia | 2015 | 0.591 |

*Table 4: countries chosen from Europe*

| Column A: Country | Column B: Year of data that I will be collecting from | Column C: Gini coefficient |
|---|---|---|
| 21. United Kingdom | 2016 | 0.348 |
| 22. Slovenia | 2017 | 0.242 |
| 23. Germany | 2016 | 0.319 |
| 24. Iceland | 2015 | 0.268 |
| 25. Belgium | 2017 | 0.274 |

*Table 5: countries chosen from North America and the Caribbean*

| Column A: Country | Column B: Year of data that I will be collecting from | Column C: Gini coefficient |
|---|---|---|
| 26. United States | 2016 | 0.411 |
| 27. Guatemala | 2014 | 0.483 |
| 28. Costa Rica | 2018 | 0.480 |
| 29. Canada | 2017 | 0.333 |
| 30. Dominican Republic | 2018 | 0.437 |

## Construct the Lorenz curve by the income distributions

To work out the Gini coefficient, it is important that we construct the Lorenz curve, which in this investigation, I will draw the Lorenz curve by the data collected from THE WORLD BANK (figure 1). I will have to divide the x-axis into different intervals provided by the data from the website. As I have selected 30 countries, I will use China as an example, the other 29 countries' income distribution and Gini coefficient will be shown in Table 6.1 using the same method.

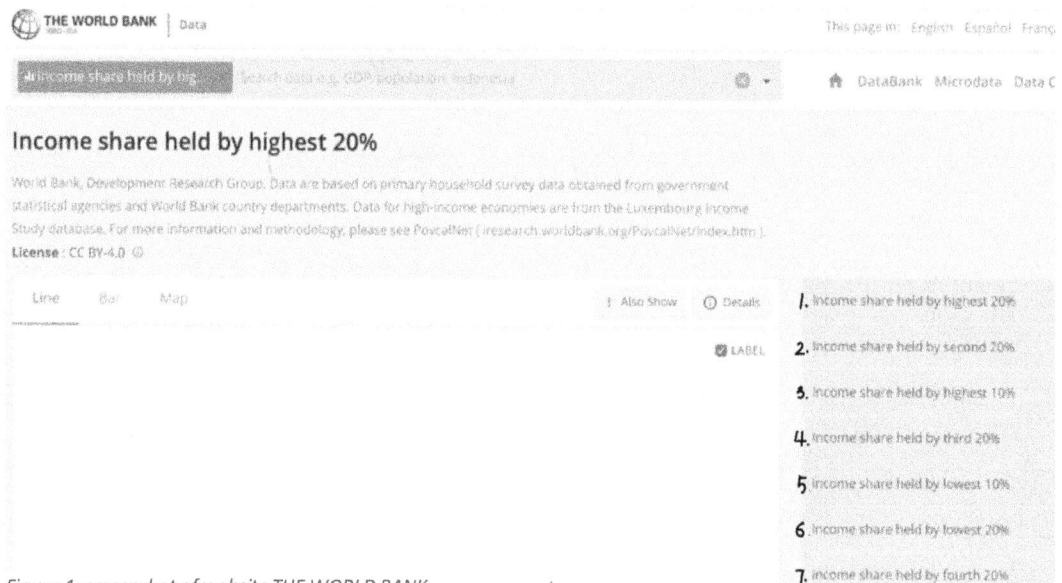

Figure 1: screenshot of website THE WORLD BANK

The data for the income distribution is split into 7 intervals,

1. Income shared by $80 \le x \le 100$

2. Income shared by $20 \le x \le 40$

3. Income shared by $90 \le x \le 100$

4. Income shared by $40 \le x \le 60$

5. Income shared by $0 \le x \le 10$

6. Income shared by $0 \le x \le 20$

7. Income shared by $60 \le x \le 80$

$x$= the cumulative population percentage from the poorest to the richest

From this data, I calculated the income shared by $10 \le x \le 20$ from 5 and 6, and $80 \le x \le 90$ from 1 and 3. Hence, the interval for income distribution of the population can be split into:

$$intervals\ of\ 10 \begin{cases} 0 \le x < 10 \\ 10 \le x < 20 \end{cases}$$

58

$$intervals\ of\ 20 \begin{cases} 20 \leq x < 40 \\ 40 \leq x < 60 \\ 60 \leq x < 80 \end{cases}$$

$$intervals\ of\ 10 \begin{cases} 80 \leq x < 90 \\ 90 \leq x \leq 100 \end{cases}$$

The income distribution shared percentage for China is shown in table 6 and other 29 countries in table 6.1 (appendix). To construct the curve, both of my axes are starting from the origin and maximum up to 100, no value can exceed that as the Lorenz curve is a cumulative graph. To draw the graph, I need the cumulative values from the 2nd row taken from table 6, and they are made into another table 7 (values for other 29 countries in table 7.2) and all values added up from column b to h (table 6 and 6.1) should always be 100. However, some might not be due to the rounding at each stage. While plotting the values on graph, the y values are the cumulative values on 2nd row in table 7.

*Table 6: Column b to h shows the different intervals of percentage of population ordered by income*

*2nd row shows percentage of income shared by the percentage of population*

| Column a: country | Column b: $0 \leq x \leq 10$ | Column c: $10 \leq x \leq 20$ | Column d: $20 \leq x \leq 40$ | Column e: $40 \leq x \leq 60$ | Column f: $60 \leq x \leq 80$ | Column g: $80 \leq x \leq 90$ | Column h: $90 \leq x \leq 100$ |
|---|---|---|---|---|---|---|---|
| 1. China | 2.7 | 3.8 | 10.7 | 15.3 | 22.2 | 16.0 | 29.3 |

*Table 7: the cumulative percentage of income shared by the percentage of population for China*

| Row 1: cumulative values of $x$ / Column 1: countries | Column 2: $x \leq 10$ | Column 3: $x \leq 20$ | Column 4: $x \leq 40$ | Column 5: $x \leq 60$ | Column 6: $x \leq 80$ | Column 7: $x \leq 90$ | Column 8: $x \leq 100$ |
|---|---|---|---|---|---|---|---|
| 1. China | 2.7 | 6.5 | 17.2 | 32.5 | 54.7 | 70.7 | 100 |

The Lorenz curve I drew for China shown by graph 2. The total area under $y = x$ (like area A+B in graph 1) is always $= \frac{100 \times 100}{2} = 5000$. To calculate the Gini coefficient, I need to work out A on graph 3, which I can subtract area B (under $y = f(x)$) from area under $y = x$, therefore I will need to integrate $y = f(x)$ to work out the area B. However, I can't integrate the curve directly as I don't know the function, the Lorenz curved line I drew is just an approximation or the best fit line, therefore I will use numerical approaches like the trapezoidal rule and the Simpson's rule to work out the area under curve.

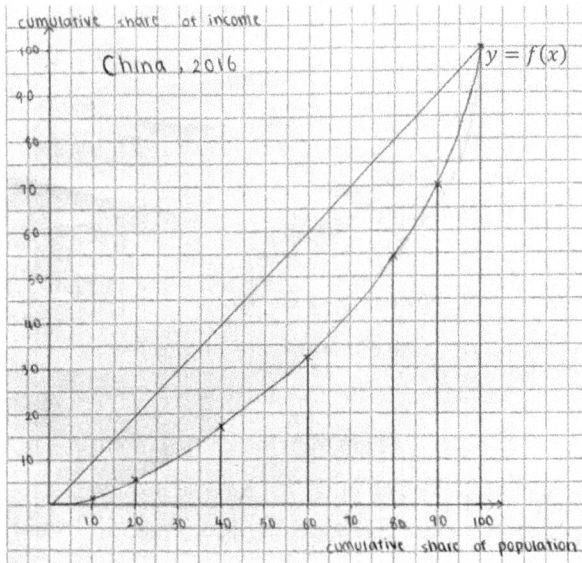

Graph 2: the Lorenz curve for China in 2016

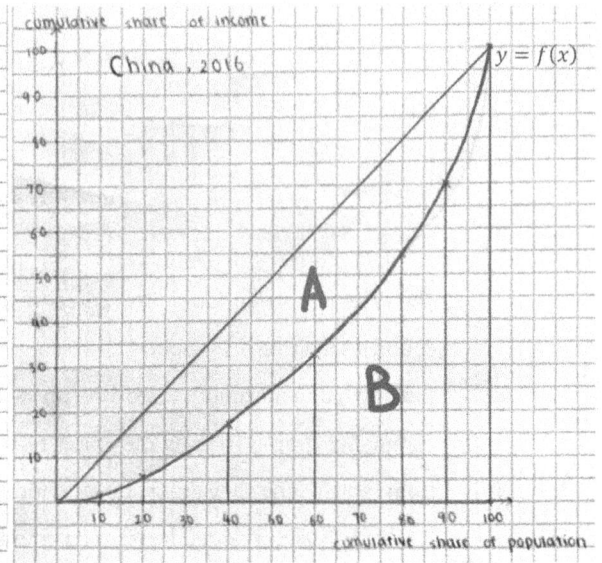

Graph 3: the area under perfect equality line split into 2 parts

Trapezoidal rule

Example of Trapezoidal rule[5]

Here is a cruve $y = f(x)$ *(graph 4)*, to get the approximation of $\int_a^b f(x)dx$, we need to first divide the domain $[a, b]$ into same interval width ($\varDelta x$), then draw a straight line between the coordinates and it forms a trapezium.

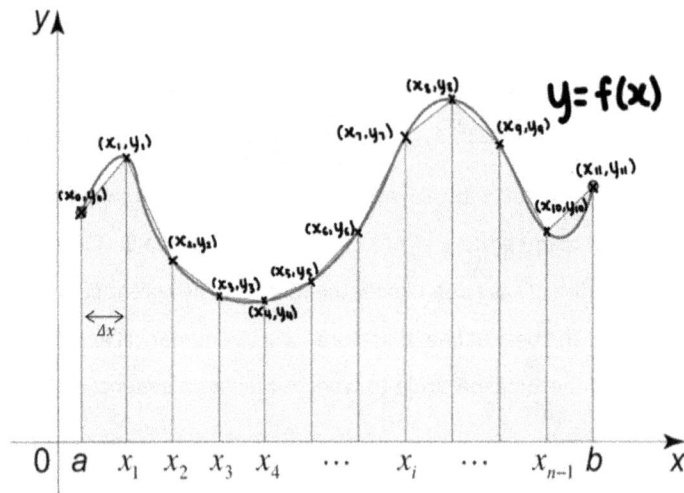

Graph 4: example of numerical integration by trapezoidal rule

---

[5] https://www.math24.net/trapezoidal-rule/

60

Height of trapzium is always $\Delta x$, which can be worked out as

$$\Delta x = \frac{upper\ boundary\ (b) - lower\ boundary\ (a)}{n}$$

Where $n$ is the the number of intervals the curve is split into (in this case is 11 equal strips)

The parallel sides of the trapezium are the $y$ coordinates, therefore the sum of the area of the trapeziums would be:

$$= \frac{\Delta x}{2}(y_0 + y_1) + \frac{\Delta x}{2}(y_1 + y_2) + \frac{\Delta x}{2}(y_2 + y_3) + \cdots + \frac{\Delta x}{2}(y_9 + y_{10}) + \frac{\Delta x}{2}(y_{10} + y_{11})$$

(where $\frac{\Delta x}{2}(y_n + y_{n+1})$ is the formula for a trapezium, derived from trapzium's general area

formula: $\frac{height}{2}(sum\ of\ parallel\ sides)$)

$$= \frac{\Delta x}{2}(y_0 + y_1 + y_1 + y_2 + y_2 + y_3 + \cdots + y_9 + y_{10} + y_{10} + y_{11})$$

$$= \frac{\Delta x}{2}(y_0 + 2y_1 + 2y_2 + 2y_3 + \cdots + 2y_{10} + y_{11})$$

Therefore the general formula for the trapezoidal rule is:

$$\int_a^b f(x)dx \approx \frac{\Delta x}{2}[f(x_0) + 2f(x_1) + 2f(x_2) + 2f(x_3) + \cdots + 2f(x_{n-1}) + f(x_n)]$$

$f(x)$ = the hypothetical function of the cruve

$f(x_n)$ = the y coordinate

$\Delta x$ = the interval (width of the strips) and will become the height of the trapezium

($\approx$ is used as this only shows an approximation for the area)

## Application of Trapezoidal rule for China

I will split the x-axis into 5 equal strips according to data I collected from THE WORLD BANK in table 6, with each interval $\Delta x$ of 20% shown on graph 5.

$$\Delta x = \frac{100-0}{5} = 20 \text{ (width of strip)}$$

The lower and upper boundaries is 0 and 100.

The values for f(x$_n$), which are the income distribution shared percentage, are from table 7.

By substituting the values from tabel 7 into the general formula, total area is:

$$= \frac{20}{2}[0 + 2(6.5) + 2(17.2) + 2(32.5) + 2(54.7) + 100]$$

$$= 10 \times 321.8$$

$$\int_0^{100} f(x)dx \approx 3218$$

61

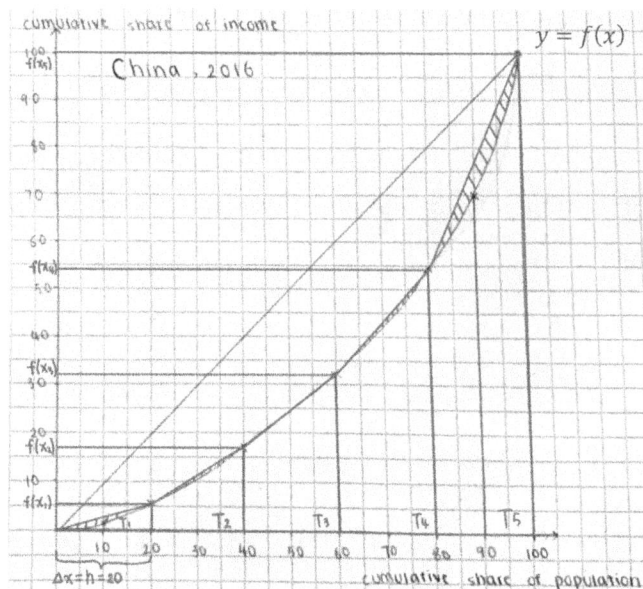

*Graph 5: the Lorenz curve for China in 2016 split into 5 equal intervals and trapezium*

This can also be proven by calculating the area of each trapezium ($T_1$ to $T_5$).

Area for $T_1$

$= \dfrac{\Delta x}{2}[f(x_0) + f(x_1)]$

$= \dfrac{20}{2}(0 + 6.5)$

$= 65$

Area for $T_2$

$= \dfrac{\Delta x}{2}[f(x_1) + f(x_2)]$

$= \dfrac{20}{2}(6.5 + 17.2)$

$= 10 \times 23.7$

$= 237$

Area for $T_3$

$= \dfrac{\Delta x}{2}[f(x_2) + f(x_3)]$

$= \dfrac{20}{2}(17.2 + 32.5)$

$= 10 \times 49.7$

$= 497$

Area for $T_4$

$= \dfrac{\Delta x}{2}[f(x_3) + f(x_4)]$

$= \dfrac{20}{2}(32.5 + 54.7)$

$= 10 \times 87.2$

$= 872$

Area for $T_5$

$= \dfrac{\Delta x}{2}[f(x_4) + f(x_5)]$

$= \dfrac{20}{2}(54.7 + 100)$

$= 10 \times 154.7$

$= 1547$

$$= 65 + 237 + 497 + 872 + 1547 = 3218$$

However, the trapezoidal rule only works when the intervals $\Delta x$ are the same, which is this case is 20. Hence area between $y = x$ and $y = f(x)$ equals $5000 - 3218 = 1782$

$$Gini\ coefficient = \frac{1782}{5000}$$

$$= 0.356\ (3\ s.f)$$

I compared this result with the official data from THE WORLD BANK (see table 1) 0.385, it has a percentage error of $100\left(\dfrac{0.385-0.356}{0.385}\right) = 7.53\%\ (3\ s.f)$, less than the actual value.

This is because on graph 5, the area shaded (red diagonal strips) is the part where it exceeds the actual curve $y = f(x)$, this means that the actual value for the area between $y = x$ and $y = f(x)$ should be less than what I have calculated. Hence if area under $y = f(x)$ is larger (overestimated), the numerator is less in the Gini coefficient calculation, therefore Gini coefficient would be less than expected offcial data.

Develop trapezoidal rule further

To improve the accuracy and reduce the percentage error, as the $n \rightarrow \infty$ (number of strip that $y = f(x)$ will be split into), there will be infite number of trapziums, each with small area that exceeds $y = f(x)$, therefore the calculations will be more accurate when intervals($\Delta x$) are smaller. Since I have interval ($\Delta x$) of 10s and 20s (Table 7), I constructed another graph(5.1). The area shaded (blue) which will be the overestimated part, is much less than graph 5. However, since the intervals are not equal, I need to split the ones that has an interval of 20 (trapezium $T_3$ to $T_5$ on graph 5.1) into 10 by using approximated value on the graph, which I will be taking the mean. For example, y coordinate for $x < 30$ will be the mean of $x < 20$ and $x < 40$ (see table 7 for specific values).

$$= \frac{6.5 + 17.2}{2} = 11.85 \, (not \ rounded \ for \ accuracy)$$

I made a new table (7.1) with $\Delta x$ =10 and constructed a new graph, $f(x_0)$ is always 0.

*Table 7.1: $f(x_n)$ shows the cumulative values of the percentage of income shared by the percentage of population for 30 countries*

|  | $f(x_1)$: $x \leq 10$ | $f(x_2)$: $x \leq 20$ | $f(x_3)$: $x \leq 30$ | $f(x_4)$: $x \leq 40$ | $f(x_5)$: $x \leq 50$ | $f(x_6)$: $x \leq 60$ | $f(x_7)$: $x \leq 70$ | $f(x_8)$: $x \leq 80$ | $f(x_9)$: $x \leq 90$ | $f(x_{10})$: $x \leq 100$ |
|---|---|---|---|---|---|---|---|---|---|---|
| 1. China | 2.7 | 6.5 | 11.85 | 17.2 | 24.85 | 32.5 | 43.6 | 54.7 | 70.7 | 100 |

I chose to use the mean because the cumulative values in between the intervals that had a width of 20 (Such as $x \leq 30$) is very close to the original Lorenz curve ($y = f(x)$). Another graph is drawn when all $\Delta x = 10$ (Graph 5.2).

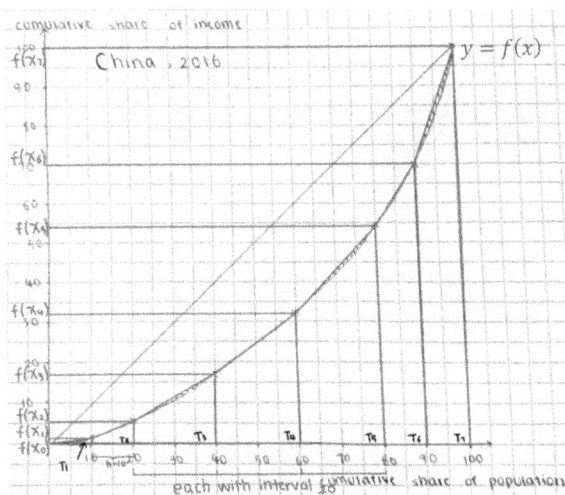

Graph 5.1: the Lorenz curve for China in 2016, intervals split unevenly

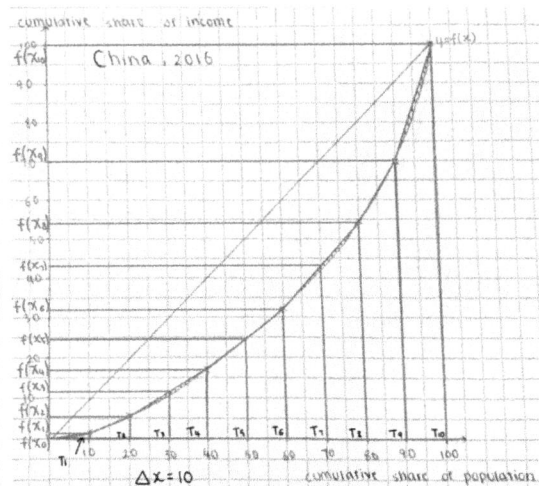

Graph 5.2: the Lorenz curve for China in 2016, intervals split evenly ($\Delta x = 10$)

Apply the general formula for trapezoidal rule to integrate area under $y = f(x)$

$$\int_0^{100} f(x)dx \approx \frac{10}{2}[0 + 2(2.7) + 2(6.5) + 2(11.85) + 2(17.2) + 2(24.85) + 2(32.5) + 2(43.6)$$

$$+ 2(54.7) + 2(70.7) + 100]$$

$$\approx 5(629.2)$$

$$\approx 3146$$

$$The\ Gini\ coefficient = \frac{5000 - 3146}{5000} = \frac{1854}{5000}$$

$$= 0.371\ (3\ s.f.)$$

So now as the intervals became smaller, the value for the Gini coefficient from the developed trapezoidal rule is even closer with the actual Gini coefficient, with a percentage error of $100\left(\frac{0.385 - 0.371}{0.385}\right) = 3.64\%\ (3\ s.f.)$.

## Simpson's rule

### Example of Simpson's rule[6]

The area under the curve is still overestimated because the trapezium's sides are all straight. Therefore, I will try to use the Simpson's rule, which the coordinates are connected with parabolas. However, to apply this rule, the number of intervals $n$ must be even.

$$\Delta x = \frac{b - a}{n} = h$$

b is the upper boundary and a is the lower boundary

$$\frac{n}{2} \in \mathbb{Z}^+, n \neq 0$$

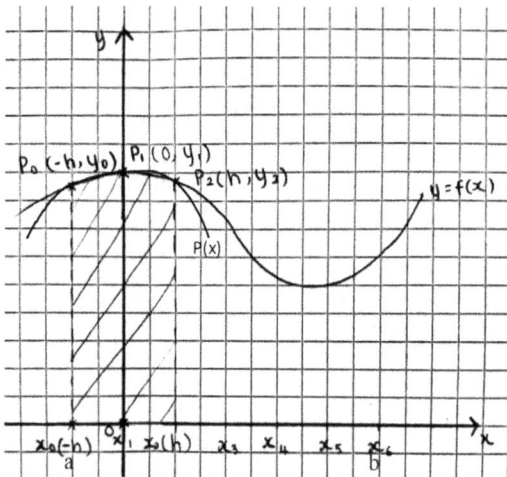

Graph 6: example of numerical integration by Simpson's rule

To approximate $\int_a^b f(x)dx$ (graph 6), we can first focus on the domain from $-h$ to $h$, which the points on the curve will be crossed through by a parabola($P(x)$) and we will give it a general form of: $y = Ax^2 + Bx + C$

And by integrating $P(x)$, we get

$$\int_{-h}^h P(x)dx = \int_{-h}^h (Ax^2 + Bx + C)dx$$

$$= \int_{-h}^h (Ax^2 + C)dx$$

Because $Bx$ is a liner odd function, it's inverted and symmetrical on the x-axis, so the integration of it equals to

[6] https://www.youtube.com/watch?v=7MoRzPObRf0

$$= \left[\frac{Ax^3}{3} + Cx\right]_{-h}^{h} = 2\left[\frac{Ax^3}{3} + Cx\right]_{0}^{h}$$

> Because $Ax^2 + C$ is an even function (symmetrical on y-axis), the integration of -h to h would be twice from 0 to h

Now I will substitute the values into $x$

$$2\left[\left(\frac{Ah^3}{3} + Ch\right) - (0 - 0)\right]$$

$$= \frac{2Ah^3}{3} + 2Ch$$

$$\int_{-h}^{h}(Ax^2 + Bx + C)dx = \frac{h}{3}(2Ah^2 + 6C)$$

We can also express the y coordinates (graph 6) by substituting $x$ value into general

parabola form:

$y_0 = A(-h)^2 - Bh + C$

$y_1 = C$

$y_2 = A(h)^2 + Bh + C$

I noticed that if we add $y_0$ and $y_1$, we get $2A(h)^2 + 2C$, it's really similar to the result from

the integration. Therefore, if we use $4\,y_1 = 4C$, we can get the result in the parenthesis.

$\int_{-h}^{h}(Ax^2 + Bx + C)dx$ can be also written as $\frac{h}{3}(y_0 + 4y_1 + y_2)$, and this is the basic

formula of the Simpson's rule. To integrate other parts of $y = f(x)$ (graph 6), we use same

method for $[x_2, x_4]$ and $[x_4, x_6]$, it would have the same basic formula as it is always a

parabola (symmetrical) when it's drawn through the points on $y = f(x)$ and Simpson's rule

integrates area under the parabola. Therefore, the sum of the area under the parabolas is:

$$\frac{h}{3}(y_0 + 4y_1 + y_2) + \frac{h}{3}(y_2 + 4y_3 + y_4) + \frac{h}{3}(y_4 + 4y_3 + y_6)$$

At this point I noticed that except from the first ($y_0$)and last y coordinate($y_n$), for $y_i$ ($i$ can

be any integer), if $i$ is an odd number, the coefficient is 4 whereas it's an even number, it

would be 2.

General formula for $y = f(x)$ (graph 6)$= \frac{h}{3}(y_0 + 4y_1 + 2y_2 + 4y_3 + 2y_4 + 4y_5 + y_6)$

The general formula for the Simpson's rule is:

$$\int_{a}^{b}f(x)dx \approx \frac{\Delta x}{3}[f(x_0) + 4f(x_1) + 2f(x_2) + 4f(x_3) + \cdots 2f(x_{n-2}) + 4f(x_{n-1}) + f(x_n)]$$

## Application of Simpson's rule for China

Using the Simpson's rule, area under $y = f(x)$ for China's Lorenz curve is

$$= \int_0^{100} f(x)dx \approx \frac{10}{3}[0 + 4(2.7) + 2(6.5) + 4(11.85) + 2(17.2) + 4(24.85) + 2(32.5) + 4(43.6) \\ + 2(54.7) + 4(70.7) + 100]$$

$$\approx \frac{10}{3}[10.8 + 13 + 47.4 + 34.4 + 99.4 + 65 + 174.4 + 109.4 + 282.8 + 100]$$

$$\approx \frac{10}{3}(936.6)$$

$$\approx 3122$$

$$The\ Gini\ coefficient = \frac{5000 - 3122}{5000} = \frac{1878}{5000}$$

$$= 0.376\ (3\ s.f.)$$

If I compare this result with the actual Gini coefficient of China, it has a percentage error of $100\left(\frac{0.385-0.376}{0.385}\right) = 2.34\%\ (3\ s.f)$, the error is reduced to a relatively low number, although it is also quite similar to result gained from trapezoidal rule. I will use the Simpson's rule for the other 29 countries to work out their Gini coefficient. All the results including the percentage error will be in table 7.2 (appendix).

The mean of the percentage error for the Simpson's rule is 2.24 % (all values in percentage error column added and divided by 30 countries). However, for the 4th country (Indonesia), the value for percentage error is a negative number, which means the estimated Gini coefficient is higher than the real one from THE WORLD BANK. This might be caused by the gradient for a part of the Lorenz curve being steeper or flatter than other parts, so an anomaly can occur. However, when calculating the mean for percentage error, modulus will be taken as I wanted to find how far off are the Gini coefficients from their real value. Overall, the percentage error is quite low, therefore it is a reliable integrating method to use.

## Conclusion

In this exploration, the aim was to first use the income distribution data to construct the Lorenz curve. As discussed before, some countries update their data less frequent, this might be because some government don't tend to share data about their income distribution, however, as long as the percentage of income shared are from the same year, the results will still be reliable. From that, I tried to calculate the Gini coefficient by first trying to work out area under Lorenz curve using 2 different integrating methods.

The trapezoidal rule was used by splitting the intervals into 5 equal strips, and percentage error was quite big (7.53%). As the number of intervals $n \to \infty$, the area under Lorenz curve is more accurate as less area would be overestimated. Hence, I developed the rule as I approximated the values by using the mean of the domains. I also used Simpson's rule which can give an even more accurate approximation of the area under curve $y = f(x)$, because it uses parabolas to estimate the area (which the shapes of parabolas are more similar to Lorenz curve as they are curve).

The average percentage error worked out form China example from the developed trapezoidal rule was 3.64%, and with Simpson's rule was only 2.34 %. Thus, I decided to use the Simpson's rule for all the other 29 countries, and I found out the average percentage error for using the Simpson's rule when intervals have a width of 10 is 2.24%.

To conclude, I explored the different ways of approximating area under the graph and reduced the percentage error examining the relative area that is overestimated by the intervals, and therefore I used the mean to approximate the value between interval width of 20. Even though each country has different years for the data, it doesn't really affect the result at the end. However, I think it would be better to use the data from the same year for every country so there can be more comparison made and potentially reduce anomalies.

# 2. MODELING LIGHT RAYS THROUGH VECTOR MATHEMATICS

Author: Luigi Pizzolito
Moderated Mark: 17/20
Level: Math AA HL

# 1 Introduction

## 1.1 Aim of the Exploration

This investigation aims to explore how mathematics allows for realistic modelling of the behaviour of light rays in order to create realistic computer-generated images of virtual scenes. In particular, to explore the rendering of simple spheres in a 3D environment, and the use of linear gradients and surface normal representations to provide colour and shading.

## 1.2 Personal Interest

This is a significant topic of exploration as it has many uses, from art to video games to animation and convincing visual effects. The technique of ray tracing is not new and even existed before computers, however with the recent advancement of hardware-accelerated ray tracing in modern graphics cards: we are approaching an era where real-time ray tracing is becoming more and more viable.

Image 1: Photorealistic ray-traced image (Tran, Janke, & Veledan, 2006).

Allowing it to overcome its overwhelming computation cost in trade-off for realism (see Image 1), in comparison to other rendering techniques.

Personally, I take interest in exploring the way that computer graphics has changed over generations; and ray tracing is an epitome of an application where mathematics accurately models the natural world. Additionally, ray tracing has applications outside of graphics, as it can be used to simulate other physical phenomena such as sound particles/waves. The task of programming and exploring such a ray tracing engine requires in-depth understanding of the mathematical workings of ray tracing, while also provides a fulfilling personal challenge.

## 1.3 Method

In order to carry out this exploration; ray tracing concepts will be explored by implementing them and creating a ray tracing engine using Java Script ES6, the programming language that I am most comfortable in for such an object-oriented exploration (Singh, 2020). Whereby doing so, each element of ray tracing mathematics can clearly demonstrate its role in producing realistic light behaviour within the output image, in a step-by-step manner. Which I believe can bring the most value and understanding to this mathematical exploration; by contrasting the mathematics with the image it creates.

# 2 Body

## 2.1 Essence of Ray Tracing

In essence, ray tracing is a 3D rendering technique that works by simulating the path of light rays realistically to generate computer graphics, implementing light phenomena and behaviour that occurs within real light.

In order to do so, a 'virtual' camera is setup with a point (the viewpoint) and a plane ahead of that point (the viewport). The viewport is split up into square pixels of an image, and a light ray is cast from the viewpoint to each individual pixel. The path that this ray takes as it travels from the viewpoint, through the viewport, to the scene, is then used to calculate intersections with objects and other interactions; to ultimately determine the pixel colour for that ray. When repeated for all the pixel, this creates a ray traced image ("Ray tracing (graphics)", 2001) (Cross, 2013).

The objects in the scene are defined mathematically, but they also have material properties just like real objects; which include properties such as colour, diffusion, reflectiveness, roughness etc. It is this combination of simulating individual light rays with realistic light behaviour, and realistic modelling of materials and objects, that gives ray tracing it's photorealism.

It is advised that the reader be familiar with the concept of iteration, as all of the calculations and most of the variables have to be iteratively repeated/recalculated for each individual pixel in order to construct a final output image. Additionally, one may notice that this process is reversed when compared to reality; the light rays emanate from the camera to the scene, rather than from the scene towards the camera. This is deliberately done to reduce the number of rays that need to be simulated to a fixed number of rays per pixel; visually, reversing this process has no impact.

### 2.1.1 Vectors & Rays

In order to simplify the mathematics to handle the interactions between 3D geometry, vectors are used to provide a framework. 3D vectors are defined by a set of three real numbers, which define the magnitude and direction of the vector. This can be visualised as the three real values that define $x, y$ & $z$ coordinates of a point which the vector points to from the origin (0,0,0). They are represented by bold variables and the components are written in a vertical square bracket:

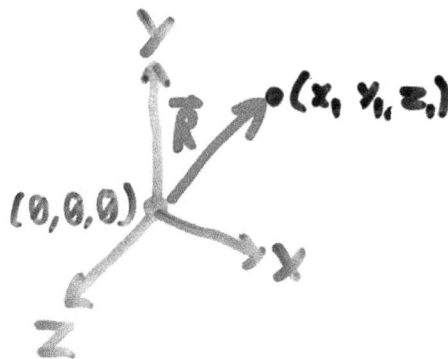

Diagram 1: XYZ axis and a vector as defined between the origin and (x, y, z).

71

$$\vec{R} = \begin{bmatrix} x \\ x \\ z \end{bmatrix}$$

$$\vec{R} \in \mathbb{R}^3, \qquad x, y, z \in \mathbb{R}$$

By using 3D vectors to define, manipulate and store elements, the mathematics becomes a lot more straightforward and the benefits of being able to use operations such as dot or cross products and vector addition/subtraction are gained.

### Rays

To define rays, two vectors can be used; $A$ defines the point which the vector originates from and $\vec{B}$ defines the direction of the ray. Giving the ray equation below:

$$P(t) = A + t\vec{B}$$

$$P, A, \vec{B} \in \mathbb{R}^3, \qquad t \in \mathbb{R}$$

Where $P$ is a point on the ray at $t$. This equation is also the vector equation of a line.

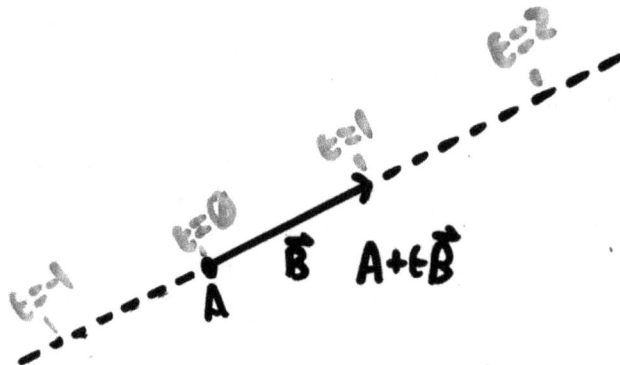

Diagram 2: Ray/line equation, showing how $t$ multiplies the direction vector $\vec{B}$.

In some sense, $A$ is the origin of the ray and $t$ is a factor of the direction $\vec{B}$ in which the ray has travelled.

When it is said that a ray is cast, what is often meant is that a ray is established at point $A$ and then $t$ is increased in the direction $\vec{B}$ until an event such as an intersection occurs. The act of 'casting' rays is analogous to the way that light rays travel in reality.

### 2.1.2 Camera & Casting Rays

Let us create a camera (shown below), in order to start casting rays:

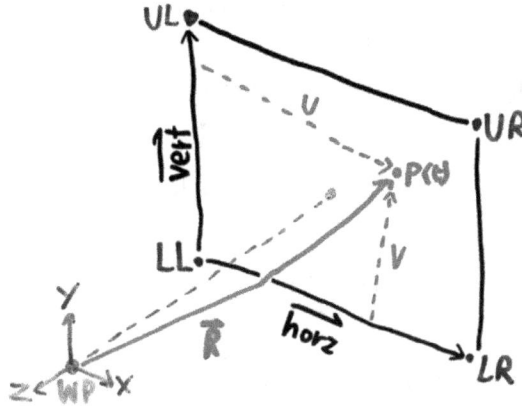

Diagram 3: Setup of camera; with viewpoint, viewport and ray casting.

As previously mentioned, to create an image rays are cast from viewpoint **WP** through a point on the rectangular viewport (the rectangle created by **UL, UR, LR, LL**), shown in Diagram 3. To calculate the point **P**($t$) which lays within the viewport, the vertices of the viewport are placed at a fixed focal length in the $Z$- direction (dashed green line) from the viewpoint **WP** and spaced in such a way that creates a rectangle (**UL, UR, LR, LL**) which is centered on **WP**.

The viewport is split up into pixels in width and height. By iterating over these pixels; a ratio of the current width and height progressed ($u, v$) can be used to scale horizontal and vertical vectors ($\overrightarrow{\boldsymbol{Vert}}, \overrightarrow{\boldsymbol{Horz}}$) from the bottom left corner (**LL**) of the viewport rectangle in order to find the point **P**($t$). The ray is then created, originating at the viewpoint (**WP**) and in the direction of point **P**($t$), through which the ray is cast to interact with objects in the scene to determine the colour of that pixel. (Hollasch, 2020).

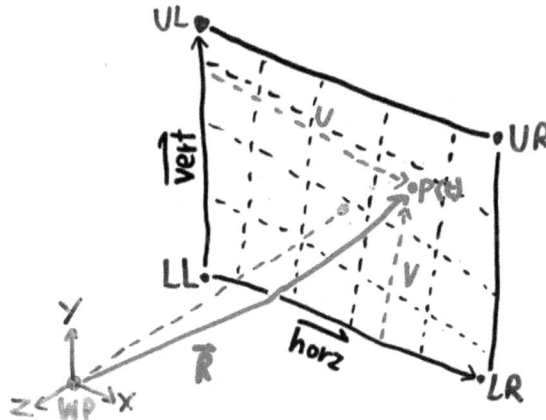

Diagram 4: Camera diagram with viewport split into pixel subdivisions.

## Worked Example

First, we calculate the points that define the viewpoint and viewport, according to the diagrams above.

For simplicity, let the viewpoint, $\boldsymbol{WP}$, equal the origin $(0, 0, 0)$,

let the viewport, have a width of 4 and height of 2,

let the focal length (green line) be 1.

This creates a camera placed at the origin, with an aspect ratio of 2:1. With this information we can calculate the four points $\boldsymbol{UL}$, $\boldsymbol{UR}$, $\boldsymbol{LR}$ and $\boldsymbol{LL}$.

$$\text{Corner Point} = WP + \begin{bmatrix} \pm \dfrac{WP_{\text{width}}}{2} \\ \pm \dfrac{WP_{\text{height}}}{2} \\ -WP_{\text{focal length}} \end{bmatrix}$$

$$\boldsymbol{UL} \qquad \qquad \boldsymbol{UR} \qquad \qquad \boldsymbol{LR} \qquad \qquad \boldsymbol{LL}$$

$$= \begin{bmatrix} 0 \\ 0 \\ 0 \end{bmatrix} + \begin{bmatrix} -2 \\ +1 \\ -1 \end{bmatrix} \quad = \begin{bmatrix} 0 \\ 0 \\ 0 \end{bmatrix} + \begin{bmatrix} +2 \\ +1 \\ -1 \end{bmatrix} \quad = \begin{bmatrix} 0 \\ 0 \\ 0 \end{bmatrix} + \begin{bmatrix} +2 \\ -1 \\ -1 \end{bmatrix} \quad = \begin{bmatrix} 0 \\ 0 \\ 0 \end{bmatrix} + \begin{bmatrix} -2 \\ -1 \\ -1 \end{bmatrix}$$

$$= \begin{bmatrix} -2 \\ 1 \\ -1 \end{bmatrix} \qquad \quad = \begin{bmatrix} 2 \\ 1 \\ -1 \end{bmatrix} \qquad \quad = \begin{bmatrix} 2 \\ -1 \\ -1 \end{bmatrix} \qquad \quad = \begin{bmatrix} -2 \\ -1 \\ -1 \end{bmatrix}$$

This then allows us to calculate our vertical and horizontal scaling vectors, $\overrightarrow{\boldsymbol{Vert}}$ and $\overrightarrow{\boldsymbol{Horz}}$ respectively, which is done through vector subtraction.

$$\overrightarrow{\boldsymbol{Vert}} = \boldsymbol{UL} - \boldsymbol{LL} \qquad\qquad\qquad \overrightarrow{\boldsymbol{Horz}} = \boldsymbol{LR} - \boldsymbol{LL}$$

$$= \begin{bmatrix} -2 \\ 1 \\ -1 \end{bmatrix} - \begin{bmatrix} -2 \\ -1 \\ -1 \end{bmatrix} \qquad\qquad = \begin{bmatrix} 2 \\ -1 \\ -1 \end{bmatrix} - \begin{bmatrix} -2 \\ -1 \\ -1 \end{bmatrix}$$

$$= \begin{bmatrix} 0 \\ 2 \\ 0 \end{bmatrix} \qquad\qquad\qquad\qquad = \begin{bmatrix} 4 \\ 0 \\ 0 \end{bmatrix}$$

With those points calculated, point $\boldsymbol{P}(t)$ can be calculated to construct the ray. One ray is iteratively cast for each pixel of the output image.

Below is a step-by-step example of how the rays $\vec{R}$, are constructed. In this case, let the ray go through pixel $(5, 20)$, given that pixel $(0, 0)$ is at $\boldsymbol{LL}$, let the image be 50 by 25 pixels.

| Step | | Formulas & Calculations | |
|---|---|---|---|
| 1 | Calculate the scaling factors $u$ & $v$. By finding the percentage position of the current pixel. | $u = \dfrac{\text{Pixel}_x}{\text{Image}_{\text{width}}}$ $= \dfrac{5}{50}$ $= 0.1$ | $v = \dfrac{\text{Pixel}_y}{\text{Image}_{\text{height}}}$ $= \dfrac{20}{25}$ $= 0.8$ |
| 2 | Scale the scaling vectors $\overrightarrow{\boldsymbol{Vert}}$ & $\overrightarrow{\boldsymbol{Horz}}$ by $u$ & $v$ and add the resulting vector to $\boldsymbol{LL}$, to find $\boldsymbol{P}(t)$. | $\boldsymbol{P}(t) = \boldsymbol{LL} + \left(u \times \overrightarrow{\boldsymbol{Horz}}\right) + \left(v \times \overrightarrow{\boldsymbol{Vert}}\right)$ $= \begin{bmatrix} -2 \\ -1 \\ -1 \end{bmatrix} + \left(0.1 \times \begin{bmatrix} 4 \\ 0 \\ 0 \end{bmatrix}\right) + \left(0.8 \times \begin{bmatrix} 0 \\ 2 \\ 0 \end{bmatrix}\right)$ $= \begin{bmatrix} -2 \\ -1 \\ -1 \end{bmatrix} + \begin{bmatrix} 0.4 \\ 0 \\ 0 \end{bmatrix} + \begin{bmatrix} 0 \\ 1.6 \\ 0 \end{bmatrix}$ $= \begin{bmatrix} -1.6 \\ 0.6 \\ -1 \end{bmatrix}$ | |
| 3 | Calculate the direction vector of the ray. Hence calculate the ray from $\boldsymbol{WP}$ to $\boldsymbol{P}(t)$. | $\vec{R} = A + t\vec{B}$ $A = \boldsymbol{WP} = \begin{bmatrix} 0 \\ 0 \\ 0 \end{bmatrix}$ $\vec{B} = \overrightarrow{\boldsymbol{P}(t) - \boldsymbol{WP}}$ $= \begin{bmatrix} -1.6 \\ 0.6 \\ -1 \end{bmatrix} - \begin{bmatrix} 0 \\ 0 \\ 0 \end{bmatrix}$ $= \begin{bmatrix} -1.6 \\ 0.6 \\ -1 \end{bmatrix}$ $\therefore \vec{R} = \begin{bmatrix} 0 \\ 0 \\ 0 \end{bmatrix} + \begin{bmatrix} -1.6 \\ 0.6 \\ -1 \end{bmatrix} t$ | |

A ray $\vec{R}$ is iteratively calculated for every pixel on the image. To determine the pixel colour, the ray is extended/cast past the viewport until it either never intersects anything or it intersects with an object in the scene. Objects are defined as vector equations themselves so that they can be equated to the ray equation, allowing for solutions to be found. These solutions are in terms of $t$ and define the point along the ray where the intersection occurs. For this investigation, in the case that no intersections occur, a gradient background is used for the colour.

## Colours as Vectors and a Gradient Background

To represent colours mathematically, one must know that in computers colours are stored by their components of red, green, and blue light. This is handy as vectors are matrices with three values, which allows us to reuse vectors in order to store and manipulate colour values.

$$C = \begin{bmatrix} r \\ g \\ b \end{bmatrix}$$

One caveat is that with 8-bit colour as is commonplace with most monitors, the values for each component must be integers ranging from 0 to 255 or $2^8$ possible values. Which gives colour vectors the following limits:

$$\{0,0,0\} \leq \{C_r, C_g, C_b\} < \{2^8, 2^8, 2^8\}, \qquad C \in \mathbb{Z}^{+3}$$

Observing that the components of $C$ must be positive integers, we can define some mathematical notation for rounding in order to convert our real number answers from our calculations into integers; for this we can merge the bracket notation for celling brackets (rounding up): $\lceil x \rceil$, with those of floor brackets (rounding down): $\lfloor x \rfloor$ in order to create a rounding bracket;

$$\lfloor x \rceil$$

This notation of storing colours as vectors along with using rounding to generate integers allows us to use any other vector operations on colour. For example, to average colours $C_1, C_2, C_3, C_4$ we can just compute the following:

$$C_{\text{Avg}} = \left\lfloor \frac{C_1 + C_2 + C_3 + C_4}{4} \right\rceil$$

Another application of storing colours as vectors is allowing for the easy computation of linear gradients. To mix the following colours:

$$C_1 = \begin{bmatrix} \ \\ \ \\ \ \end{bmatrix}$$

$$C_2 = \begin{bmatrix} 255 \\ 255 \\ 255 \end{bmatrix}$$

The following equation can be used, where $t$ is the mixing factor, remembering to round the vector components to the nearest integer:

$$C_{\text{mixed}} = \lfloor (1-t) \cdot C_1 + t \cdot C_2 \rceil$$

$$0 \leq t \leq 1, \qquad t \in \mathbb{R}$$

Which gives us the following gradient:

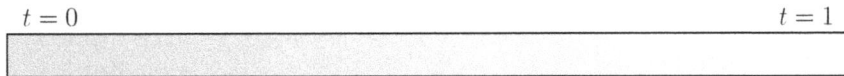

$t = 0$                                                   $t = 1$

A vertical gradient will be used for the background of the ray tracer, which is rendered when the ray undergoes no intersections.

## Worked Example

Let the gradient colours be black and magenta:

$$C_1 = \begin{bmatrix} 192 \\ 32 \\ 255 \end{bmatrix}$$

$$C_2 = \begin{bmatrix} 0 \\ 0 \\ 0 \end{bmatrix}$$

And let the mixing factor, $t$ be the $y$ component of our ray's direction vector $\vec{B}$, when scaled from its original -1 to 1 range to a range of 0 to 1. Which will give us a vertical gradient as $\vec{B}$ goes from positive to negative one along the height of the frame.

$$t = \frac{\vec{B}_y + 1}{2}$$

With that information, we can now calculate the colour value of the pixel at (5, 20).

| Step | | Formulas & Calculations |
|---|---|---|
| 1 | Using Ray $\vec{R}$, as previously calculated for pixel (5, 20) | $\vec{R} = \begin{bmatrix} 0 \\ 0 \\ 0 \end{bmatrix} + \begin{bmatrix} -1.6 \\ 0.6 \\ -1 \end{bmatrix} t$ |

77

| 2 | Calculate the gradient mixing factor, $t$ | $t = \dfrac{\vec{B}_y + 1}{2}, \vec{B} = \begin{bmatrix} -1.6 \\ 0.6 \\ -1 \end{bmatrix}$ <br><br> $= \dfrac{0.6 + 1}{2}$ <br><br> $= 0.8$ |
|---|---|---|
| 3 | Calculate the pixel colour, using the linear gradient equation. Making sure to round the components to integers. | $C_{\mathbf{mixed}} = \lfloor (1-t) \cdot \boldsymbol{C_1} + t \cdot \boldsymbol{C_2} \rceil$ <br><br> $= \left\lfloor (1-0.8) \begin{bmatrix} 192 \\ 32 \\ 255 \end{bmatrix} + 0.8 \begin{bmatrix} 0 \\ 0 \\ 0 \end{bmatrix} \right\rceil$ <br><br> $= \left\lfloor \begin{bmatrix} 38.4 \\ 6.4 \\ 51 \end{bmatrix} + \begin{bmatrix} 0 \\ 0 \\ 0 \end{bmatrix} \right\rceil$ <br><br> $= \begin{bmatrix} 38 \\ 6 \\ 51 \end{bmatrix}$ |
| | Result | |

This same process can then be repeated to calculate the colour of another pixel, say pixel (40, 5); which yields the colour: $\begin{bmatrix} 154 \\ 26 \\ 204 \end{bmatrix}$.

By iterating and calculating the colour values for all the other pixels of the image, just like shown. The final results creates the following image, Which can be used to verify our calculations (keeping in mind that automatic file compression may alter colour values to reduce file size when saving the image):

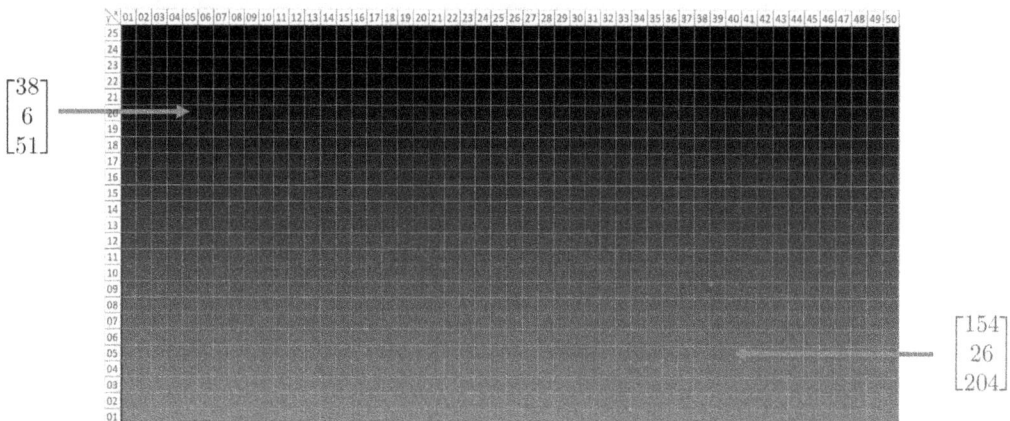

$\begin{bmatrix} 38 \\ 6 \\ 51 \end{bmatrix}$

$\begin{bmatrix} 154 \\ 26 \\ 204 \end{bmatrix}$

Render 1: A plain black to magenta gradient.

### 2.1.3 Objects (Spheres)

Next, when objects are added to our scene, the rays are extended out from the camera to check if they intersect with objects, in which case the object should be drawn instead of the background gradient.

A sphere is one of the simplest shapes to describe through an equation, which makes it ideal to begin with as our first primitive. With the cartesian system, a sphere with radius $R$, with its centre on the origin is traditionally defined with the equation shown below, where a sphere is the collection of all points that have the same distance $R$ from the origin:

$$x^2 + y^2 + z^2 = R^2$$

If the sphere's centre is not at the origin, but at say $(C_x, C_y, C_z)$, the equation becomes:

$$(x - C_x)^2 + (y - C_y)^2 + (z - C_z)^2 = R^2$$

To convert this equation into vector form, we can define any point on the sphere as $P = (x, y, z)$ and the sphere's centre at $C = (C_x, C_y, C_z)$. This allows us to write the vector from the centre of the sphere to a point on the surface as $(\vec{P - C})$, illustrated below:

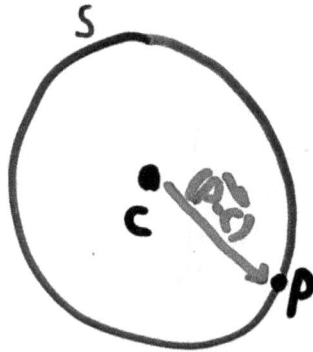

Diagram 5: Sphere $S$; defined by its center point and radius using vectors.

By then squaring this term by multiplying itself with the dot product, we obtain the expression below, which expands to the expression previously shown above:

$$(\vec{P - C}) \cdot (\vec{P - C}) = (x - C_x)^2 + (y - C_y)^2 + (z - C_z)^2$$

Where the dot product is defined as the scalar sum of each component in one vector multiplied by the equivalent component in the other vector. For example: $(x_1, y_1, z_1) \cdot (x_2, y_2, z_2) = x_1 x_2 + y_1 y_2 + z_1 z_2$

With that established, we can derive the vector equation of a sphere by simply equating our expression to $R^2$, via substitution of $(\vec{P - C}) \cdot (\vec{P - C})$ back into the original sphere formula for a sphere at centre $C$:

$$(\vec{P - C}) \cdot (\vec{P - C}) = R^2$$

In order to then check for intersections between the ray cast by our camera viewpoint, we can now substitute our ray equation, $\boldsymbol{P}(t) = \boldsymbol{A} + t\vec{\boldsymbol{B}}$ into the point at the sphere's surface $\boldsymbol{P}$. Which yields the following:

$$\left(\overline{\boldsymbol{P}(t) - \boldsymbol{C}}\right) \cdot \left(\overline{\boldsymbol{P}(t) - \boldsymbol{C}}\right) = R^2$$

$$\left(\overline{\boldsymbol{A} + t\vec{\boldsymbol{B}} - \boldsymbol{C}}\right) \cdot \left(\overline{\boldsymbol{A} + t\vec{\boldsymbol{B}} - \boldsymbol{C}}\right) = R^2$$

Rearranging and expanding:

$$\left(\overline{t\vec{\boldsymbol{B}} + \boldsymbol{A} - \boldsymbol{C}}\right) \cdot \left(\overline{t\vec{\boldsymbol{B}} + \boldsymbol{A} - \boldsymbol{C}}\right) = R^2$$

$$(\vec{\boldsymbol{B}} \cdot \vec{\boldsymbol{B}})t^2 + (\vec{\boldsymbol{B}} \cdot (\overline{\boldsymbol{A} - \boldsymbol{C}}))2t + \left(\left((\overline{\boldsymbol{A} - \boldsymbol{C}}) \cdot (\overline{\boldsymbol{A} - \boldsymbol{C}})\right) - R^2\right) = 0$$

This gives us a $2^{\text{nd}}$ degree polynomial whose roots are values of $t$ where the ray intersects the sphere. Which has the following possibilities:

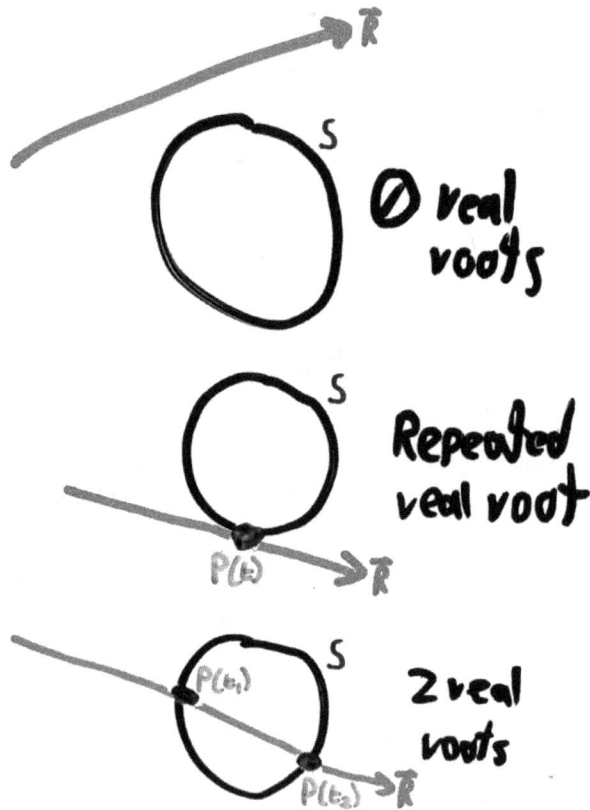

Diagram 6: Three possibilities for solutions when a ray intersects a sphere.

Furthermore, since these solutions can be calculated using the quadratic equation; to save on compute time the discriminant can be calculated first to determine the number of solutions in order to avoid attempting to calculate non-real solutions when the ray does not intersect.

Additionally, to simplify the calculation even more, one can notice that the second coefficient of the polynomial for the $t$ term has a factor of 2; this allows for a substitution of $b = 2b_{\frac{1}{2}}$ in order to eliminate that factor with a simplified version of the quadratic formula:

$$ t = \frac{-b \pm \sqrt{b^2 - 4ac}}{2a} $$

$$ t = \frac{-2b_{\frac{1}{2}} \pm \sqrt{\left(2b_{\frac{1}{2}}\right)^2 - 4ac}}{2a} $$

$$ t = \frac{-2b_{\frac{1}{2}} \pm 2\sqrt{b_{\frac{1}{2}}^2 - 4ac}}{2a} $$

$$ t = \frac{-b_{\frac{1}{2}} \pm \sqrt{b_{\frac{1}{2}}^2 - 4ac}}{a} $$

By using this formula instead, we can compute using half of the second coefficient ($b_{\frac{1}{2}}$) and improve our render times by subsequently performing fewer operations to calculate the determinant and the point of intersection. Which is required to calculate the pixel's colour and summon other recursive rays.

Side note: When there are two real roots, the smallest one (closest to the viewpoint) is chosen to determine the intersection; this is the version of the quadratic formula where the root of the determinant is subtracted.

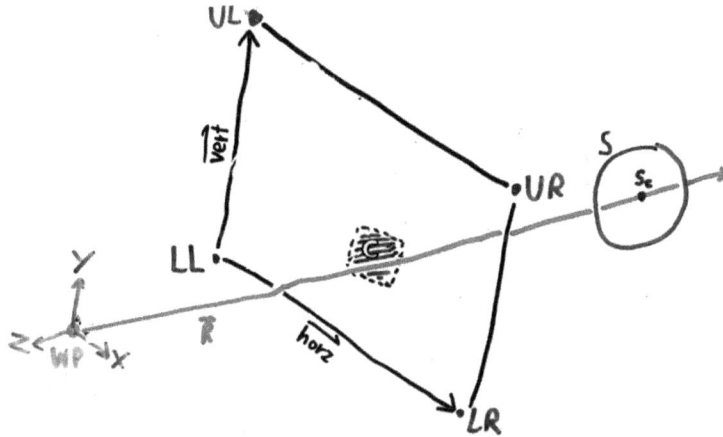

Diagram 7: Color of a viewport pixel is determined based on object intersections.

**Worked Example**

Let sphere $S$, at point $C = \begin{bmatrix} 0 \\ 0 \\ -1 \end{bmatrix}$ and radius $R = 0.5$ be defined by:

$$(\boldsymbol{P}(t) - \boldsymbol{C}) \cdot (\boldsymbol{P}(t) - \boldsymbol{C}) = R^2$$

$$\left( \boldsymbol{P}(t) - \begin{bmatrix} 0 \\ 0 \\ -1 \end{bmatrix} \right) \cdot \left( \boldsymbol{P}(t) - \begin{bmatrix} 0 \\ 0 \\ -1 \end{bmatrix} \right) = 0.5^2$$

We can then calculate the discriminant and solutions for $t$, and return a different colour, say pure red <u>if</u> an intersection occurs (discriminant $>0$).

$$C_{intersec.} = \begin{bmatrix} 255 \\ 0 \\ 0 \end{bmatrix}$$

The following notation is used to define an if statement:

$$x = \text{condition} \begin{cases} \text{value if true} \\ \text{value if false} \end{cases}$$

Let us subsequently calculate the pixel value for pixel (30, 10).

| Step | | Formulas & Calculations |
|---|---|---|
| 1 | Using Ray $\vec{R}$, as calculated for pixel (30, 10) | $\vec{R} = \boldsymbol{A} + t\vec{B} = \begin{bmatrix} 0 \\ 0 \\ 0 \end{bmatrix} + \begin{bmatrix} 0.4 \\ -0.2 \\ -1 \end{bmatrix} t$ |
| 2 | Calculate the quadratic coefficients: $a, b_{\frac{1}{2}}$ & $c$. | $a = \vec{B} \cdot \vec{B}$ <br><br> $= \begin{bmatrix} 0.4 \\ -0.2 \\ -1 \end{bmatrix} \cdot \begin{bmatrix} 0.4 \\ -0.2 \\ -1 \end{bmatrix}$ <br><br> $= (0.4 \times 0.4) + (-0.2 \times -0.2) + (-1 \times -1)$ <br><br> $= 0.16 + 0.04 + 1$ <br><br> $= 1.2$ |

| | | |
|---|---|---|
| | | $b_{\frac{1}{2}} = \vec{B} \cdot (A - C)$ $= \begin{bmatrix} 0.4 \\ -0.2 \\ -1 \end{bmatrix} \cdot \left( \begin{bmatrix} 0 \\ 0 \\ 0 \end{bmatrix} - \begin{bmatrix} 0 \\ 0 \\ -1 \end{bmatrix} \right)$ $= \begin{bmatrix} 0.4 \\ -0.2 \\ -1 \end{bmatrix} \cdot \begin{bmatrix} 0 \\ 0 \\ 1 \end{bmatrix}$ $= (0.4 \times 0) + (-0.2 \times 0) + (-1 \times 1)$ $= -1$ |
| | | $c = \left( (A - C) \cdot (A - C) - R^2 \right)$ $= \left( \left( \begin{bmatrix} 0 \\ 0 \\ 0 \end{bmatrix} - \begin{bmatrix} 0 \\ 0 \\ -1 \end{bmatrix} \right) \cdot \left( \begin{bmatrix} 0 \\ 0 \\ 0 \end{bmatrix} - \begin{bmatrix} 0 \\ 0 \\ -1 \end{bmatrix} \right) \right) - 0.5^2$ $= \left( \begin{bmatrix} 0 \\ 0 \\ 1 \end{bmatrix} \cdot \begin{bmatrix} 0 \\ 0 \\ 1 \end{bmatrix} \right) - 0.25$ $= \left( (0 \times 0) + (0 \times 0) + (1 \times 1) \right) - 0.25$ $= 1 - 0.25$ $= 0.75$ |
| 3 | Calculate the discriminant. | $\Delta = b_{\frac{1}{2}}{}^2 - ac$ $= (-1)^2 - (1.2 \times 0.75)$ $= 0.1$ |
| 4 | Determine if the gradient background (from prior calculations) or $C_{intersec.}$ should be drawn. If there is an intersection, calculate the point where it occurs, by finding the smallest value of $t$ (solution closest to camera). | $\mathbf{C} = \Delta \geq 0 \begin{cases} C_{intersec.} \\ C_{mixed} \end{cases}$ $\Delta = 0.1$ $\therefore$ Intersection occurs. Intersection point can be found. $t = \frac{-b_{\frac{1}{2}} \pm \sqrt{\Delta}}{a}$ $t = \frac{1 - \sqrt{0.1}}{1.2}$ $t = 0.570$ |

<table>
<tr><td></td><td>

$$P(t) = \vec{R}(t) = A + t\vec{B} = \begin{bmatrix} 0 \\ 0 \\ 0 \end{bmatrix} + \begin{bmatrix} 0.4 \\ -0.2 \\ -1 \end{bmatrix} t$$

$$P(0.570) = \begin{bmatrix} 0 \\ 0 \\ 0 \end{bmatrix} + \begin{bmatrix} 0.4 \\ -0.2 \\ -1 \end{bmatrix} \times 0.570$$

$$= \begin{bmatrix} 0 \\ 0 \\ 0 \end{bmatrix} + \begin{bmatrix} 0.228 \\ -0.114 \\ -0.570 \end{bmatrix}$$

$$= \begin{bmatrix} 0.228 \\ -0.114 \\ -0.570 \end{bmatrix}$$

$$\therefore C = C_{intersec.}$$

$$= \begin{bmatrix} 255 \\ 0 \\ 0 \end{bmatrix}$$

</td></tr>
<tr><td>Result</td><td></td></tr>
</table>

We can then repeat this for another pixel, say pixel (5, 20), which does not intersect, and returns the colour $\begin{bmatrix} 38 \\ 6 \\ 51 \end{bmatrix}$ from the background gradient.

Iteratively completing these calculations for every pixel then returns the following image, allowing us to check our manually calculated pixels:

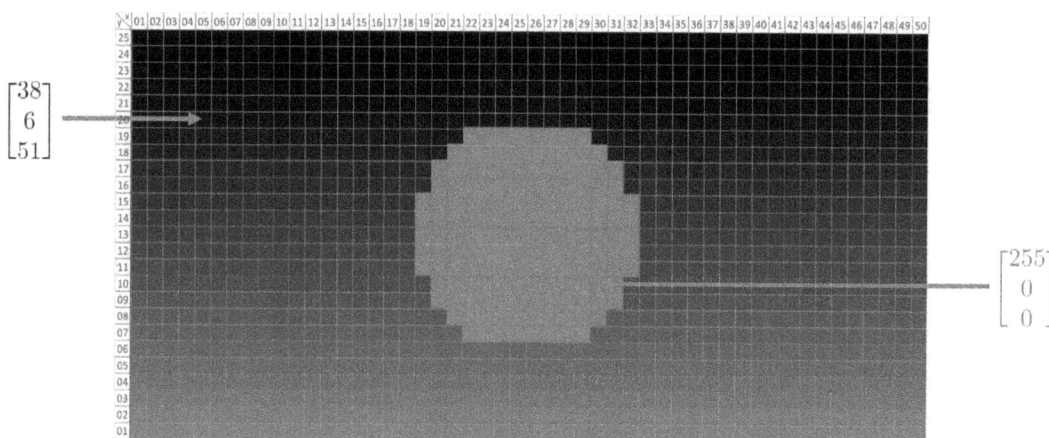

Render 2: Checking for intersections with a sphere and setting intersecting pixels to have a red colour.

## 2.2 Normal Shading

Another important vector that must be calculated in order to later compute the directions of diffused, reflected and refracted rays is the surface normal where an intersection or 'hit' occurs. The surface normal is a unit vector that is perpendicular to the surface of the object which the ray intersected.

To calculate the surface normal of a sphere we can take advantage of vector subtraction, as the normal is the same vector that is returned when the point on the surface of the sphere is subtracted from the sphere's centre point.

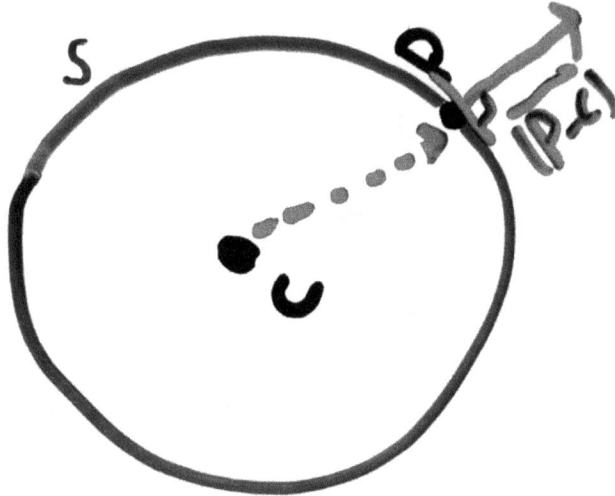

Diagram 8: Using the vector between the center point and a point on the surface of a sphere to calculate the surface normal vector.

Therefore, all that is left to do is to normalise the vector into a unit vector by dividing each of its components by its magnitude:

$$\vec{N} = \overrightarrow{P(t) - C}$$

$$\overrightarrow{N_{\text{norm}}} = \frac{\overrightarrow{P(t) - C}}{\left|\overrightarrow{P(t) - C}\right|}$$

One additional aspect that can be tracked given the surface normal is whether the ray is intersecting the outside surface of the sphere or the inside surface of the sphere. In the case that the ray is intersecting the inside surface of the sphere, the surface normal must be inverted accordingly to ensure that it faces in towards the centre of the sphere rather than away from it. This is important to later calculate refraction when ray tracing transparent and translucent spheres, as with rules such as Snell's law; the angle change as a ray changes physical medium (from ball material to air or vice versa) is dependent on whether the ray is incoming or exiting a surface.

One simple way of determining whether the normal should be inverted is to simply take the dot product of the ray's direction vector $\vec{B}$ and the normal vector $\overrightarrow{N_{\text{Norm}}}$. A negative result indicates that the vectors are in the same direction and therefore the normal does not need to be inverted. A positive result indicates that the vectors are in opposite directions and the normal should be inverted.

However, as refraction will not be implemented within this investigation, we may overlook and ignore this step of inversing normal from a ray that intersects on an inner surface of the sphere. It will be therefore assumed that all surface normals are outwards facing for this investigation.

**Worked Example**

In order to provide the sphere rendered with plain red in the previous example with some pseudo-shading, the surface normal of the intersection point in each pixel can be calculated, and the components of $\overrightarrow{N_{\text{norm}}}$ can be scaled from the real range of -1 to 1 to the integer range of 0 to 255 in order to visualise the surface vectors as R, G & B colours respectively.

Calculating the normal vector and colour for pixel (30, 10)

| Step | | Formulas & Calculations |
|---|---|---|
| 1 | Using Ray $\vec{R}$, sphere with $\mathbf{C}$ and $\mathbf{R}$, the discriminant and the intersection point at $t$, as calculated for pixel (30, 10) | $\vec{R} = \mathbf{A} + t\vec{B} = \begin{bmatrix} 0 \\ 0 \\ 0 \end{bmatrix} + \begin{bmatrix} 0.4 \\ -0.2 \\ -1 \end{bmatrix} t$ <br><br> $\mathbf{C} = \begin{bmatrix} 0 \\ 0 \\ -1 \end{bmatrix}, \qquad R = 0.5$ <br><br> $\Delta = 0.1, \qquad t = 0.570$ <br><br> $\mathbf{P}(0.570) = \begin{bmatrix} 0.228 \\ -0.114 \\ -0.570 \end{bmatrix}$ |
| 2 | Calculating the normal vector $\vec{N}$ | $\vec{N} = \overrightarrow{\mathbf{P}(t) - \mathbf{C}}$ <br><br> $= \begin{bmatrix} 0.228 \\ -0.114 \\ -0.570 \end{bmatrix} - \begin{bmatrix} 0 \\ 0 \\ -1 \end{bmatrix}$ <br><br> $= \begin{bmatrix} 0.228 \\ -0.114 \\ 0.43 \end{bmatrix}$ |

| | | |
|---|---|---|
| 3 | Normalising the normal vector to $\overrightarrow{N_{\text{norm}}}$ | $$\overrightarrow{N_{\text{norm}}} = \frac{\overrightarrow{P(t) - C}}{\left|\overrightarrow{P(t) - C}\right|}$$ $$= \frac{\begin{bmatrix} 0.228 \\ -0.114 \\ 0.43 \end{bmatrix}}{\left\|\begin{bmatrix} 0.228 \\ -0.114 \\ 0.43 \end{bmatrix}\right\|}$$ $$= \frac{\begin{bmatrix} 0.228 \\ -0.114 \\ 0.43 \end{bmatrix}}{\sqrt{0.228^2 + (-0.114)^2 + 0.43^2}}$$ $$= \frac{\begin{bmatrix} 0.228 \\ -0.114 \\ 0.43 \end{bmatrix}}{0.5}$$ $$= \begin{bmatrix} 0.456 \\ -0.228 \\ 0.860 \end{bmatrix}$$ |
| 4 | Converting the normal vector to a colour representation | $$C_{\bar{N}} = \left\lfloor \frac{\overrightarrow{N_{\text{norm}}} + 1}{2} \times 255 \right\rfloor$$ $$= \left\lfloor \frac{\begin{bmatrix} 0.456 \\ -0.228 \\ 0.860 \end{bmatrix} + 1}{2} \times 255 \right\rfloor$$ $$= \left\lfloor \begin{bmatrix} 185.64 \\ 98.43 \\ 237.15 \end{bmatrix} \right\rfloor$$ $$= \begin{bmatrix} 186 \\ 98 \\ 237 \end{bmatrix}$$ |
| | Result | |

We can then repeat this for another pixel, say pixel (22, 19). Which returns the colour $\begin{bmatrix} 81 \\ 221 \\ 203 \end{bmatrix}$.

Iteratively completing these calculations for every pixel then returns the following image, once again allowing us to check our manually calculated pixels:

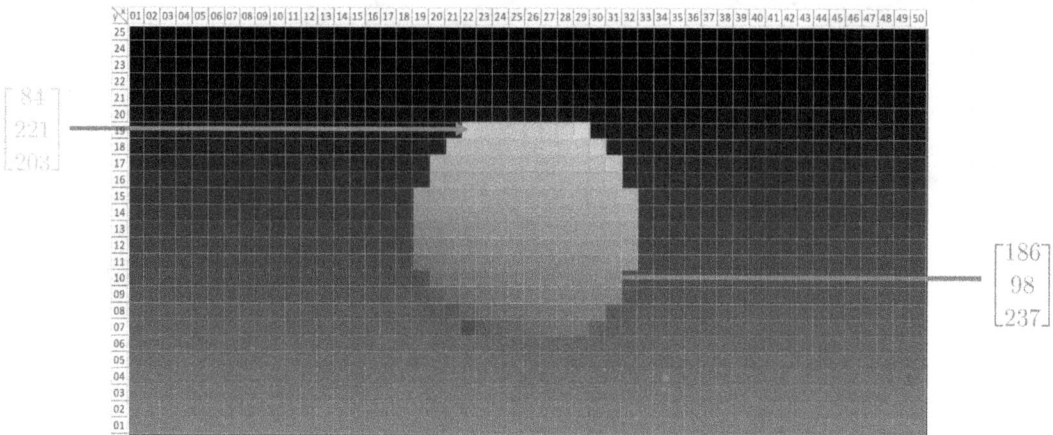

Render 3: Using a RGB representation of surface normals to provide shading.

## 2.3  Antialiasing

One of the issues of trying to quantise the continuous intersection boundaries of an object into discrete pixels is that if only a single ray is cast through the centre of the pixel the resulting image has sharp edges. That is to say that within the area of one pixel, different points within it will result in different ray paths, different intersections and subsequently different pixel colours. Antialiasing is the process of collecting multiple colour samples by casting multiple rays through random points within the area of a single pixel and then averaging the colour values returned. This removes the sharp edges and mixes the colours found at the boundary of an object.

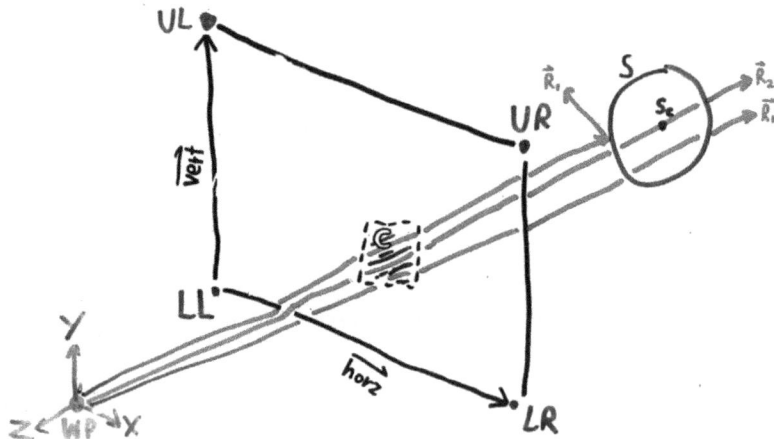

Diagram 9: Casting multiples rays per pixel to implement antialiasing.

## Worked Example

The rays are shifted by adding a random value to the scaling factors, $u$ & $v$, that is between 0 and the width of a single pixel (1). The original equations for $u$ & $v$ were:

$$u = \frac{\text{Pixel}_x}{\text{Image}_{\text{width}}} \qquad\qquad v = \frac{\text{Pixel}_y}{\text{Image}_{\text{height}}}$$

They now become the equations below, where $\xi$ is a random number:

$$\xi \in \mathbb{R}, \xi \in [0,1)$$

$$u = \frac{\text{Pixel}_x + \xi}{\text{Image}_{\text{width}}} \qquad\qquad v = \frac{\text{Pixel}_y + \xi}{\text{Image}_{\text{height}}}$$

This skews the computed ray to point to a random point within the pixel, rather than always pointing to the lower left corner of the pixel area. This shift is large enough that different colours are returned from the sampled rays, keeping within the area covered by the pixel. Each result from each sample ray is then averaged:

$$C_{\text{avg}} = \frac{\sum_{s=0}^{n_{\text{samples}}} C(s)}{n_{\text{samples}}}$$

Where: $C_{\text{avg}}$ is the anti-aliased output colour,
$n_{\text{samples}}$ is the number of samples taken,
$s$ is the current sample index,
$C(s)$ is the colour resultant from the ray, from sample $s$.

Calculating, say 4 samples/rays for each pixel, for pixel (22, 19):

| Random Number ($\xi$) | Horizontal Scaling Factor ($u$) | Random Number ($\xi$) | Vertical Scaling Factor ($v$) | Ray Equation ($\vec{R}$) | Discriminant ($\Delta$) | Pixel Colour ($C$) | |
|---|---|---|---|---|---|---|---|
| 0 | 0.440 | 0 | 0.760 | $\vec{R} = \begin{bmatrix} 0 \\ 0 \\ 0 \end{bmatrix} + \begin{bmatrix} -0.240 \\ 0.520 \\ -1 \end{bmatrix} t$ | 0.004 | $\begin{bmatrix} 84 \\ 220 \\ 203 \end{bmatrix}$ | |
| 0.128 | 0.443 | 0.843 | 0.794 | $\vec{R} = \begin{bmatrix} 0 \\ 0 \\ 0 \end{bmatrix} + \begin{bmatrix} -0.228 \\ 0.588 \\ -1 \end{bmatrix} t$ | -0.048 | $\begin{bmatrix} 42 \\ 7 \\ 56 \end{bmatrix}$ | |
| 0.975 | 0.460 | 0.368 | 0.775 | $\vec{R} = \begin{bmatrix} 0 \\ 0 \\ 0 \end{bmatrix} + \begin{bmatrix} -0.160 \\ 0.550 \\ -1 \end{bmatrix} t$ | 0.004 | $\begin{bmatrix} 110 \\ 188 \\ 17 \end{bmatrix}$ | |
| 0.341 | 0.447 | 0.569 | 0.783 | $\vec{R} = \begin{bmatrix} 0 \\ 0 \\ 0 \end{bmatrix} + \begin{bmatrix} -0.212 \\ 0.566 \\ -1 \end{bmatrix} t$ | -0.024 | $\begin{bmatrix} 42 \\ 7 \\ 55 \end{bmatrix}$ | |

Notice that not all samples even intersect the sphere (see the negative and positive discriminants), resulting in different colours. The colours computed from each of these sample rays is then averaged using the same method as described in the section **Colours as Vectors and a Gradient Background**.

$$C_{\text{avg}} = \frac{C_1 + C_2 + C_3 + C_4}{n_{\text{samples}}}$$

$$= \frac{\begin{bmatrix} 84 \\ 221 \\ 203 \end{bmatrix} + \begin{bmatrix} 42 \\ 7 \\ 56 \end{bmatrix} + \begin{bmatrix} 110 \\ 188 \\ 17 \end{bmatrix} + \begin{bmatrix} 42 \\ 7 \\ 55 \end{bmatrix}}{4}$$

$$= \begin{bmatrix} 70 \\ 106 \\ 83 \end{bmatrix}$$

When the image is rendered with 4 anti-aliasing samples, like in the worked solution above; the result is the following:

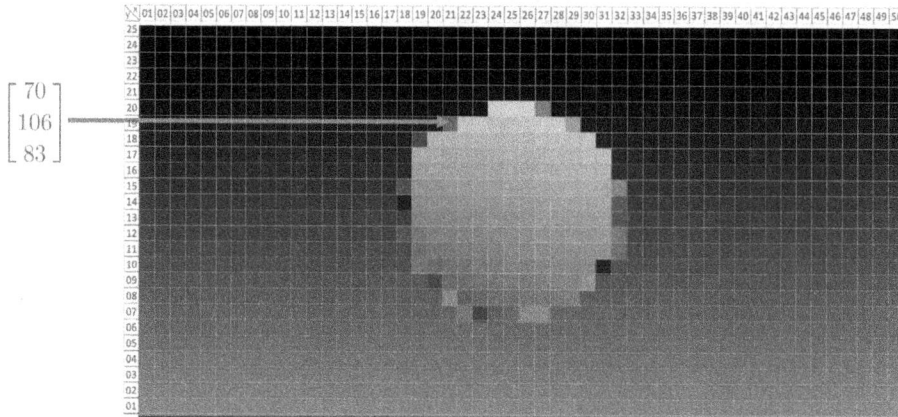

Render 4: Smoother edges of a sphere rendered with 4 sample antialising.

Comparing side-by-side with the non-anti-aliased image:

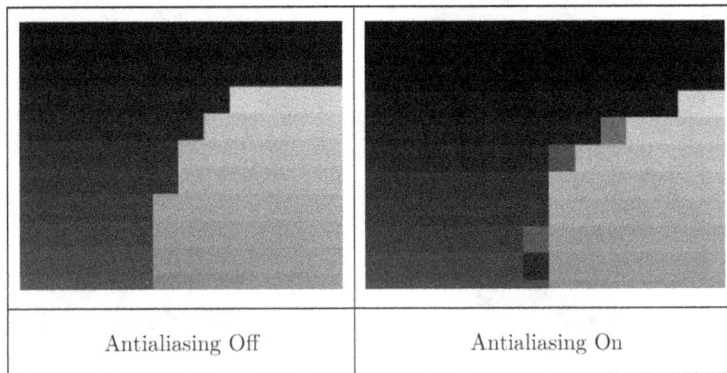

| Antialiasing Off | Antialiasing On |
| --- | --- |

Diagram 10: Close up comparison of antialiased edges.

Although this may not seem like a significant improvement at such a low resolution image, the difference becomes more evident and the antialiasing smoother if we increase the resolution of the image, for example full HD with 2000x1000 pixels:

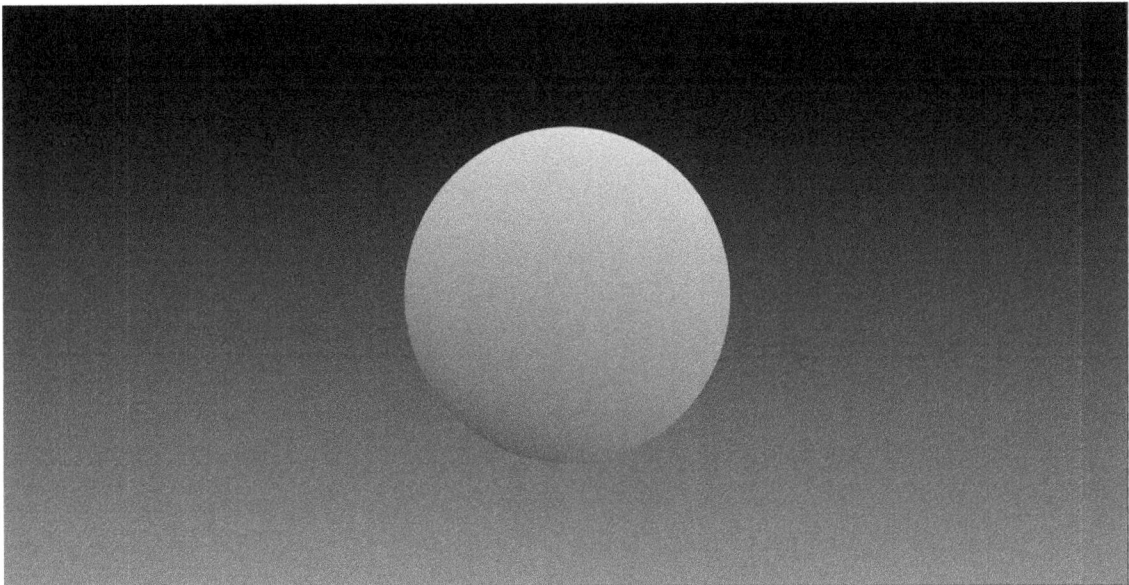

Render 5: High resolution render of a surface normal shaded sphere, containing 2,000,000 pixels.

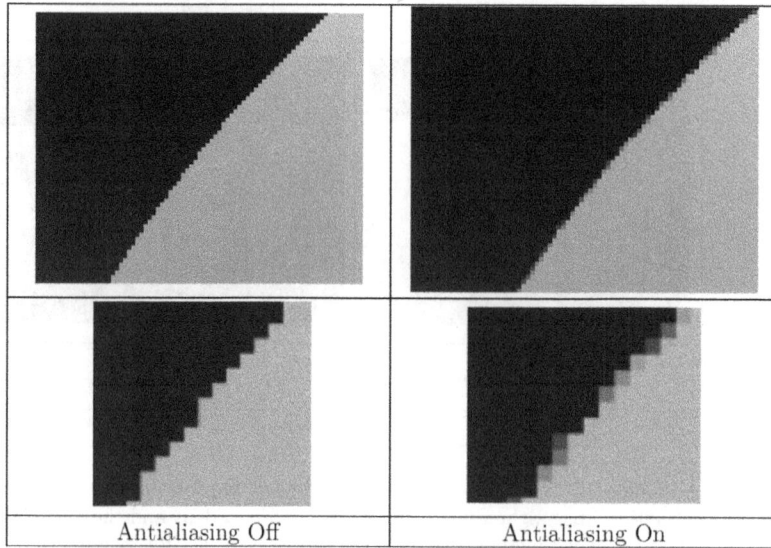

| Antialiasing Off | Antialiasing On |

Diagram 11: Close up comparison of antialiased edges in high resolution renders.

## 2.4 Recursion

So far, rays have only been cast from the camera, or viewpoint. However, in order to simulate diffusion, reflection, refraction and other light phenomena, we must be able to recursively summon more rays from an intersection point with an object. For this to occur, when an intersection happens other rays should be cast (for example reflected rays), originating at the intersection point in the calculated direction. Recursion is this process of initiating a whole new ray calculation in order to continue computing the colour of a previous ray.

A limitation of recursion (which is not implemented in this exploration) is that it adds significantly more complexity and calculation, as those summoned rays themselves may also intersect and summon child rays. This is one of the factors that makes ray tracing so computationally expensive; as we are now casting a potentially exponentially higher number of rays. To counter this, the amount or depth of recursion is usually limited, and it is common to have different recursion limits (eg. Only allow 5 ray reflections) in order to compromise between computing time and the final image quality (Hollasch, 2020).

Additional problems also arise. Currently there is only one object, one sphere in our scene, and when checking for intersections we only check for intersections with that single sphere. A prerequisite to implementing recursion would be to adapt the current method to be aware of multiple objects simultaneously (to allow the same ray to interact with all objects).

The computational complexity therefor renders a manual exploration of recursive ray tracing unviable for this mathematics-focused investigation, limiting this exploration from implementing any further light dynamics.

## 2.5 Multiple Objects

However, there is one method to attempt to get around those limitations of building an intersection checking implementation that is aware of all objects. Which is to effectively create a compositor. The role of the compositor is to calculate intersections with all objects in the scene for each pixel, but then only display the object that is closest to the camera (smallest $t$ at $\boldsymbol{P}(t)$) to the image output. This does not allow for recursion or implementation of any more advanced light dynamics, but it does allow us to add an unlimited number of objects to the scene.

Below is a render with three spheres, where the highlighted red areas are pixels in which the compositor is active and makes a choice as to which sphere should be displayed as multiple spheres intersect the ray at those pixels:

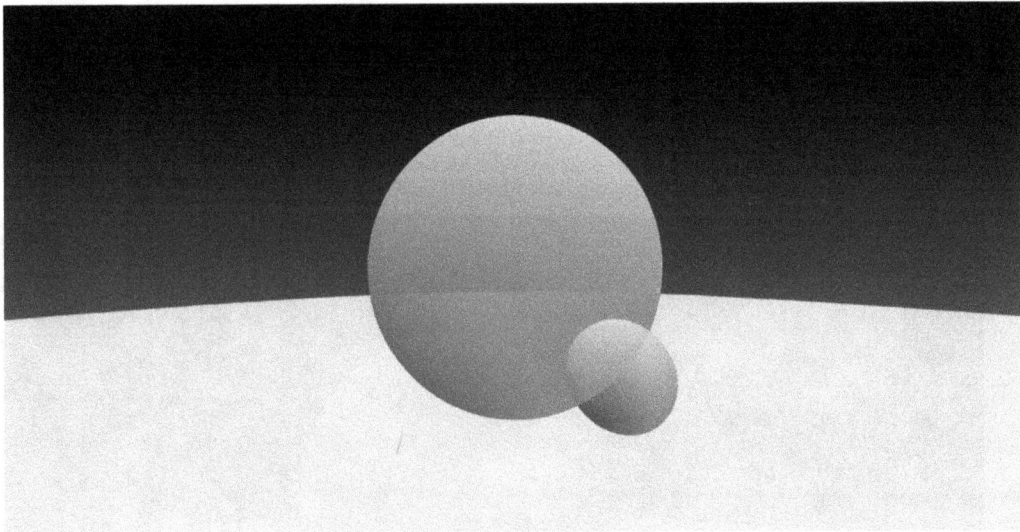

Render 6: Where red highlights where the compositor code executes.

The final result is multi-object renders like the following:

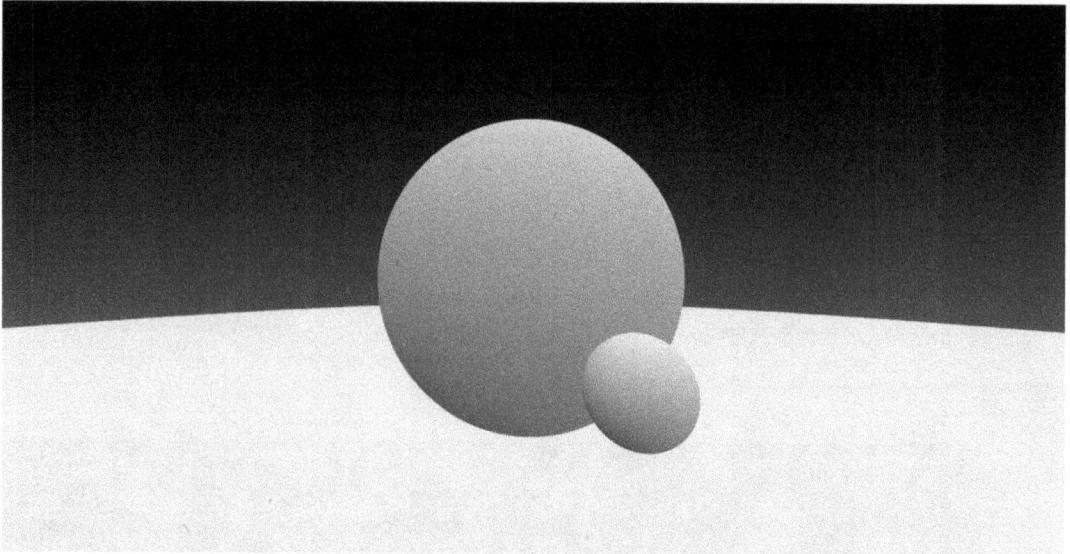

Render 7: Rendering 3 spheres.

Or adding even more sphere objects to the scene:

Render 8: Rendering seven spheres.

One thing to note is the distortion which increases as an object becomes closer to the edge of the viewport, which is a side-effect of the simple plane projecting technique used to create the camera.

# 3 Conclusion

In summary, colours, directions and the location of points were defined as 3D vectors, which allowed for the definition of rays and spheres through the use of vector equations. A camera which casts rays from a viewpoint through each pixel of a viewport placed ahead of the viewpoint was created. A gradient background was calculated by mixing two colours based on the $y$ component of the ray direction cast from the camera. Spheres were created by defining a point and a radius, which allowed for rays to be checked for intersection with spheres. This was done by solving a quadratic equation with vector components, which was computed with a modified quadratic formula; providing the amount of intersections (0, 1 or 2) and the distance/coordinates of the intersection point. Surface normal vectors were then calculated for every point on the sphere where a pixel ray is cast, allowing for pseudo-shading by mapping the normal's $X$, $Y$, & $Z$ components to the pixel's RGB colour. Antialiasing was then implemented, by collecting several randomly skewed colour samples per pixel and averaging them for the final pixel colour, eliminating sharp edges that occurred on object/background borders. Last but not least, some compositing logic was created to check distances from objects to the camera, and therefore allowing multiple objects/spheres to be rendered at once by only drawing the object nearest to the camera. Concluding by rendering such a multi-sphere scene as (Render 7 & Render 8).

Although this investigation only touches the very basics of ray tracing using only spheres as objects, without even implementing any light bounces or recursion, which would allow for the simulation of more advanced light phenomena such as diffusion, refraction, reflection and more. It has been sufficient to show the nature of ray tracing and provided a step-by-step demonstration of how each implementation of a concept affects the output render. All while keeping a concrete grasp of the vector mathematics through the calculations, despite the object-oriented computer programming, logic and variable management needed in order to complete billions of these calculations to render the full images shown.

## 3.1 Challenges

Although the mathematical concept of vectors may be simple on paper, the complexity of all the steps involved to perform the ray tracing algorithm proves to be quite counterintuitive, particularly the concept of recursion and casting rays from a camera as opposed to from a light source. The common mathematics used, such as solving quadratics, vector equations and averages, provides a strong abstraction from the task at hand: creating an image, which is a very unusual way to apply mathematics. Ray tracing continues to be counter-intuitive even once the algorithm is understood, as not only is it extremely easy to make mistakes in its implementation, but some phenomena are hard to comprehend. For instance, I was personally unable to find a method to correct the perspective distortion that occurs around the edges of the frame, which is apparent in Render 8 for example as the spheres start to look like ellipsoids.

In addition, a large challenge in the process of this exploration was performing the sheer number of calculations needed. For example, in the 2000x1000 HD renders such as Render 7; there are 2,000,000 pixels to be rendered, and our ray tracing intersection-checking colour function executes three times, once per object in the scene, giving us a minimum of 6,000,000 calculations to render that image. Needless to say, this makes ray tracing unfeasible to calculate manually, but it also introduces challenges in programming as keeping track of all the variables must be done smartly. In this case, vastly made easier through the use of object-oriented programming which is complex but allows for instancing of predefined objects making such kinds of massive, individualised calculations possible.

It is an elementary mistake to underestimate the human ability to discern that something is wrong or unrealistic with an image, even if not being able to explicitly name what. Our brains are hard wired to automatically interpret light behaviour subconsciously. This makes reaching true photorealism with ray-tracing all the more challenging, meticulous, and impressive.

## 3.2  Further Exploration

The most straightforward path to further explore ray tracing would be to continue implementing more mathematics to simulate light rays even more realistically. As ray tracing truly has the ability to generate photorealistic images. The image below was my first experience with ray tracing, rendering a scene I created in the open-source software *Blender* with the *Cycles* ray tracing engine:

Image 2: My own ray-traced image, created with *Blender* and *Cycles renderer*.

Expansions to this specific exploration could include implementing a recursive ray intersection function to allow for the rays to truly be aware of all objects in the scene, or alternatively the implementation of primitives other than spheres or ultimately the ability to render tri-tessellated based geometry which would allow for importing and rendering of any 3D object model. Additionally, other aspects such as light diffusion could be implemented by scattering light randomly when it hits a diffuse-coloured (matt) surface, or refraction by bending light rays when they travel across boundaries of physical materials; through the use of material-specific refractive indexes and Snell's law; $n_1 \sin\theta_1 = n_2 \sin\theta_2$ where $n$ is the refractive index and $\theta$ is the incident/refracted angle for each material across a boundary respectively (Hollasch, 2020). There are of course, many, many more light phenomena that should be implemented for added realism, for instance ambient occlusion, HDRI environments, subsurface scattering, chromatic aberrations, glossy, reflective, transparent & emissive materials, thin-film interference, etc, just to name a few.

There is a limitless number of features and additional mathematical algorithms within ray tracing that could be explored, especially as it is such an active, current field of research; just recently breaking through the possibility of real-time ray tracing. The limit is only in computation time/power and programming complexity.

# 4 Bibliography

Cross, D. (2013). *Fundamentals of ray tracing*. Retrieved from
http://cosinekitty.com/raytrace/raytrace_ebook.pdf

Hollasch, S. (2020). Ray tracing in one weekend series. Retrieved from
https://raytracing.github.io/

Hollasch, S. (2020). *GitHub: Ray tracing in one weekend book series*.
Retrieved January 25, 2021, from
https://github.com/RayTracing/raytracing.github.io

Ray tracing (graphics). (2001, September 27). Retrieved from
https://en.wikipedia.org/wiki/Ray_tracing_(graphics)

*Ray tracing illustration first bounce* [Diagram]. (2017). Retrieved from
https://en.wikipedia.org/wiki/File:Ray_Tracing_Illustration_First_Bounce.png

Shirley, P., Hollasch, S., & Black, T. D. (2020). *Ray tracing in one weekend*. Retrieved from
https://raytracing.github.io/books/RayTracingInOneWeekend.html

Singh, G. (2020). GitHub:Gurpreetsingh-Exe/raytracer.
Retrieved January 25, 2021, from
https://github.com/gurpreetsingh-exe/raytracer

# 3. CELESTIAL NAVIGATION

Author: Anonymous
Moderated Mark: 18/20
Level: Math AA HL

For my MYP Personal Project, I wanted to turn my passion for astronomy into a hobby I could physically do something with. I built a reflecting telescope out of repurposed parts like a part of a sewage pipe for the telescope's tube and added a computerized night-sky assistance device to make the telescope easy to use. My goal was to make astronomy accessible not only for me, but also for others who shared my interest but found it hard to involve themselves due to the lack of beginner-friendly resources. Observing the moon, planets and stars have been a great experience, but I began wondering what kind of practical application observing celestial bodies could have.

Whilst learning about the coordinate systems for celestial bodies to better understand the night sky map program my telescope used, I learnt of an example of a practical application: astronavigation, also known as celestial navigation. With relatively limited tools and an understanding of the positions of celestial bodies relative to the Earth, sailors can determine their location on Earth with little to no digital technology. Celestial navigation dates back centuries before the invention of GPS, but even with the omnipresence of GPS, it is still an essential skill that sailors around the world are required to learn. All of this is possible thanks to the use of spherical geometry and trigonometry – an alternative to Euclidean planar geometry that arguably better describes the world around us, at least for the purposes of navigation on Earth.

The purpose of this investigation will be to explore the mathematics behind how celestial navigation works and apply those concepts to find my own latitude and longitude based on observation and measurement of the celestial bodies available to me. Of course, this investigation will be limited to the resources I have access to – I am not a sailor at sea, nor do I have the highest quality of equipment that real navigators usually have access to – thus I will have to make do with alternative methods of measuring the angles of celestial bodies. To better replicate the experience of sailors using astronavigation, I will also use spherical geometry to calculate the shortest distance from my location to a given destination (which I have determined to be the largest Vietnamese city of Ho Chi Minh City) and the direction I would have to travel in to get there. I will use GPS services like Google Maps to compare my actual results with the theoretical answer. By the end of this investigation, through the surrogate hypothetical situation of "flying" to an arbitrary point in Ho Chi Minh City (taking off from my house and cruising at an inconsequential altitude), the concepts behind astronavigation and their practical applications will be made clear. First of all, unfamiliar variables and concepts must be defined to aid understanding in later parts of this investigation.

## Definitions and Variables

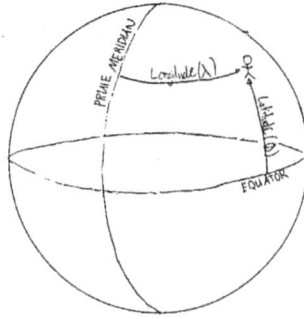

*Figure 1: Geographic Coordinate System*

**Equator:** A circle drawn at latitude 0° that is equally distant from the North Pole and the South Pole.

**Prime Meridian:** An arc drawn at longitude 0° (in Greenwich, England) that connects the North Pole and the South Pole.

| Variable | Symbol | Definition |
|----------|--------|------------|
| Latitude | $\phi$ | The angular distance from the Equator of a location on Earth. Latitudes north of the Equator are positive and latitudes south of the Equator are negative. $-90° \leq \phi \leq 90°$ |
| Longitude | $\lambda$ | The angular distance from the Prime Meridian of a location on Earth. Longitudes east of the Prime Meridian are positive and longitudes west of the Prime Meridian are negative. $-180° < \lambda < 180°$ |

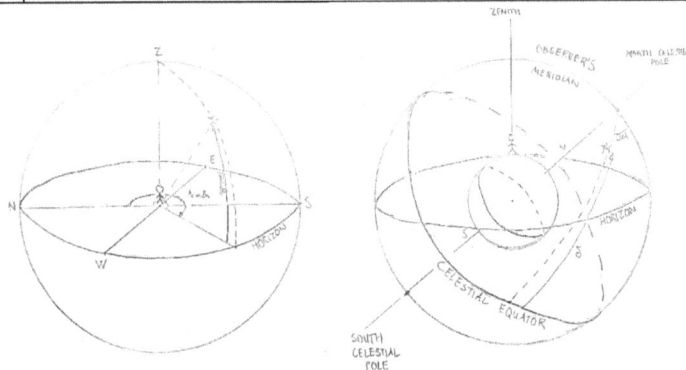

*Figure 2: The Horizontal Coordinate System (left) and the Equatorial Coordinate System (right)*

**Horizontal Coordinate System:** A coordinate system that places the cardinal directions on the plane of the observer's horizon and uses Altitude and Azimuth to determine the angular coordinates of an object. (Machtelinckx)

**Equatorial Coordinate System:** A coordinate system that is a projection of the Geographic Coordinate System onto the Celestial Sphere and uses Declination and Hour Angle determine the angular coordinates of an object. (King)

101

**Zenith:** The point that is perpendicular to the horizontal plane and is directly overhead the observer.

**Observer's Meridian:** The longitude arc of the observer projected onto the celestial sphere.

| Variable | Symbol | Definition |
|---|---|---|
| | | Horizontal Coordinate System |
| Altitude | $\alpha$ | The angular distance from the Horizon of a celestial object. Altitude is measured with a **sextant**. $0° \leq \alpha \leq 90°$ |
| Azimuth | $\gamma$ | The bearing of an object measured clockwise from north to the object's vertical circle (a great circle through the object and the zenith). Azimuth is measured with a **compass**. $0° < \gamma \leq 360°$ |
| | | Equatorial Coordinate System |
| Declination | $\delta$ | The angular distance from the Celestial Equator of a celestial object. A positive declination is north of the Celestial Equator and one that is negative is south of the Celestial Equator. $-90° \leq \delta \leq 90°$ |
| Greenwich Hour Angle | GHA | The angular distance from the Prime Meridian to a celestial object. The GHA is measured westward from the Prime Meridian to the hour circle (a great circle through the poles and the object). $0° < GHA \leq 360°$ |
| Local Hour Angle | LHA | The angular distance from the Observer's Meridian to a celestial object. The LHA is measured westward from the Observer's Meridian to the hour circle (a great circle through the poles and the object). $0° < LHA \leq 360°$ |

What I realize from defining both systems is that both coordinate systems are valid and have their purposes. The Horizontal Coordinate System is simple and both altitude and azimuth can be measured relatively easily with a sextant and a compass. The Equatorial Coordinate System is a direct counterpart to the Geographical Coordinate System, where declination is a projection of latitude and GHA is a projection of longitude. Since one is easily measurable and the other is related to the observer's geographical coordinates, connecting these systems is the key to finding my coordinates based off measurements of celestial objects.

## Great Circle Distances

Before finding my latitude and longitude, it is important that great circle distances (GCD) are understood. A GCD is the shortest distance between two points on the surface of a sphere. It is a part of a great circle, a circle with the same radius as the sphere (i.e. the Equator on Earth). The Earth is not a perfect sphere; thus these uncertainties will be addressed later. Being able to calculate a GCD is useful for calculating the distance required to go from one point to another, given that the initial and final geographical $(\phi, \lambda)$ coordinates are known (one of the goals of this investigation) (Rick). GCDs will also be important to initially finding my latitude and longitude, since spherical triangles constructed

102

from GCDs will be used to do so later in this investigation. For now, to calculate a GCD, we first consider a unit sphere with spherical radius R=1 and center O:

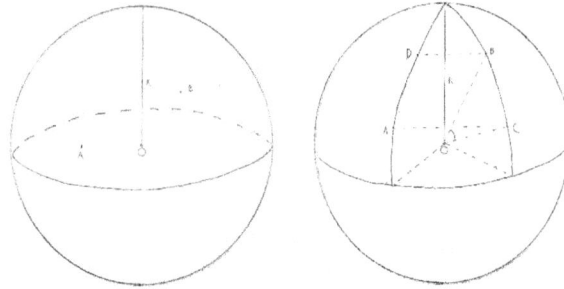

*Figure 3*

To calculate the GCD between points A ($\phi_1$, $\lambda_1$) and B ($\phi_2$, $\lambda_2$), additional points C ($\phi_1$, $\lambda_2$) and D ($\phi_2$, $\lambda_1$) must be drawn such that DBCA is a planar isosceles trapezoid. The chord AB, diagonal of the trapezoid, will allow for the angle between the two points to be found and thus the arc length or GCD to be found. We start with connecting points B and C to the center O, connecting point B to point C to form triangle OB and constructing point M as a midpoint on chord BC:

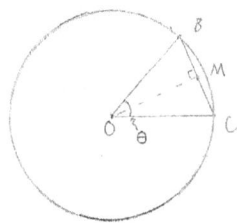

*Figure 4: Planar view of Great Circle BC*

$$\overline{OB} \cong \overline{OC} = R = 1$$

$$\angle BOC = \theta$$

$$\angle BOM = \frac{\theta}{2}$$

$$\frac{\overline{BM}}{R} = \frac{\overline{MC}}{R} = \sin\frac{\theta}{2}$$

Angle $\theta$ in triangle BOC represents the difference in latitude between the points A and B that we are interested in,

($\phi_2 - \phi_1$) or $\Delta\phi$. This is because point A is on the same latitude $\phi_1$ as point C. Since $\theta = \Delta\phi$, $\overline{BM} + \overline{MC} = \overline{BC}$, and R = 1:

$$\overline{BC} = 2R\sin\frac{\theta}{2} = 2\sin\frac{\Delta\phi}{2}$$

$$\overline{AD} \cong \overline{BC} = 2\sin\frac{\Delta\phi}{2}$$

$2r\sin\dfrac{\theta}{2}$ is also the general formula for the length of any chord given the radius of the circle r, and will later be used to the

find the chord AB.

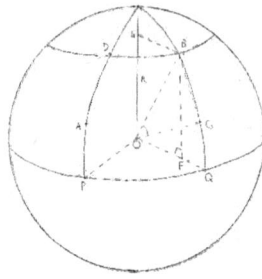

*Figure 5*

Points P $(0, \lambda_1)$ and Q $(0, \lambda_2)$ are where the lines of longitude of A and B meet the equator. The Length of chord PQ is:

$$\overline{PQ} = 2R \sin \frac{\angle POQ}{2} = 2 \sin \frac{\lambda_2 - \lambda_1}{2} = 2 \sin \frac{\Delta \lambda}{2}$$

The radius of latitude $\phi_2$ circle DB can be found by drawing BF perpendicular to OQ to form right triangle OBF and rectangle OBFG, and since OB = R = 1:

$$\overline{OF} = \cos(\angle BOF) \times \overline{OB}$$
$$\overline{OF} = \cos \phi_2 = \overline{GB}$$

With the same working we would find that the radius at latitude $\phi_1$ is $\cos\phi_1$. Substituting the radius values at latitude $\phi_2$ and $\phi_1$ as $\cos\phi_2$ and $\cos\phi_1$ respectively into the chord length formula to find chords DB and AC:

$$\overline{DB} = 2 \cos \phi_2 \sin \frac{\Delta \lambda}{2}$$
$$\overline{AC} = 2 \cos \phi_1 \sin \frac{\Delta \lambda}{2}$$

Now that we have the lengths of all the chords DB, AD, BC and AC, the next thing to do is to find the length of the chord AB. Note that the length of chord AB does not correspond with the distance between A and B, since this does not account for the curvature of the sphere. Chord AB is instead important to finding the angle between A and B and then the GCD.

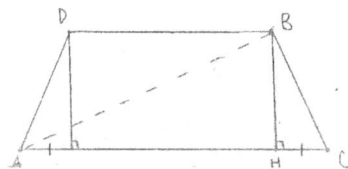

*Figure 6*

Since DBCA is an isosceles trapezoid, we have:

$$\overline{CH} = \frac{\overline{AC} - \overline{DB}}{2}$$

$$\overline{AH} = \overline{AC} - \overline{CH} = \overline{AC} - \frac{\overline{AC} - \overline{DB}}{2} = \frac{\overline{AC} + \overline{DB}}{2}$$

Using the Pythagorean theorem, we get:

$$\left(\overline{AB}\right)^2 = \left(\overline{BH}\right)^2 + \left(\overline{AH}\right)^2$$

$$\left(\overline{AB}\right)^2 = \left(\overline{BC}\right)^2 - \left(\frac{\overline{AC} - \overline{DB}}{2}\right)^2 + \left(\frac{\overline{AC} + \overline{DB}}{2}\right)^2$$

$$\left(\overline{AB}\right)^2 = \left(\overline{BC}\right)^2 + \overline{AC} \times \overline{DB}$$

The chord lengths $\overline{BC} = 2\sin\frac{\Delta\phi}{2}$; $\overline{AC} = 2\cos\phi_1 \sin\frac{\Delta\lambda}{2}$; and $\overline{DB} = 2\cos\phi_2 \sin\frac{\Delta\lambda}{2}$ can then be substituted in:

$$\left(\overline{AB}\right)^2 = 4\sin^2(\frac{\Delta\phi}{2}) + 2\cos\phi_1 \sin\frac{\Delta\lambda}{2} \times 2\cos\phi_2 \sin\frac{\Delta\lambda}{2}$$

$$\overline{AB} = 2\sqrt{\sin^2\left(\frac{\Delta\phi}{2}\right) + \cos\phi_1 \cos\phi_2 \sin^2\left(\frac{\Delta\lambda}{2}\right)}$$

And to make the next step of finding the angle $\angle AOB$ easier to understand and write, we set $a = \left(\frac{\overline{AB}}{2}\right)^2$ therefore:

$$a = \left(\frac{\overline{AB}}{2}\right)^2 = \sin^2\left(\frac{\Delta\phi}{2}\right) + \cos\phi_1 \cos\phi_2 \sin^2\left(\frac{\Delta\lambda}{2}\right)$$

To find angle $\angle AOB$ we look at isosceles triangle AOB with side lengths equal to 1:

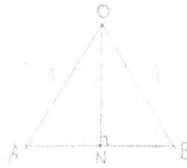

*Figure 7: Triangle AOB*

Using the Pythagorean theorem again and knowing that $\overline{AN}$ is half the length of $\overline{AB}$:

$$\overline{AN} = \frac{\overline{AB}}{2} = \sqrt{a}$$

$$\left(\overline{ON}\right)^2 = \left(\overline{AN}\right)^2 + \left(\overline{OA}\right)^2$$

$$\overline{ON} = \sqrt{1 - \left(\frac{\overline{AB}}{2}\right)^2} = \sqrt{1 - a}$$

Then finding $\angle AOB$ based on $\overline{AN}$ and $\overline{AB}$:

105

$$\tan(\angle AON) = \frac{\overline{AN}}{\overline{ON}} = \frac{\sqrt{a}}{\sqrt{1-a}}$$

$$\angle AON = \arctan\left(\frac{\sqrt{a}}{\sqrt{1-a}}\right)$$

$$\angle AOB = 2\angle AON = 2\arctan\left(\frac{\sqrt{a}}{\sqrt{1-a}}\right)$$

With angle $\angle AOB$ found, the only calculation that remains to be done to find the great circle distance is to use the arc length formula $d = R \times \theta$, where R is the radius of the great circle and $\theta$ is the angle subtended by the two points in radians. Therefore, for the formula to find the great circle distance between any two points given their latitude and longitude we have:

$$a = \sin^2\left(\frac{\Delta\phi}{2}\right) + \cos\phi_1 \cos\phi_2 \sin^2\left(\frac{\Delta\lambda}{2}\right)$$

$$\theta = 2\arctan\left(\frac{\sqrt{a}}{\sqrt{1-a}}\right)$$

$$d = R \times \theta$$

This is the Haversine Formula for finding great circle distances because the haversine is an old trigonometric ratio such that $\text{haversin}\,\theta = \sin^2\left(\frac{\theta}{2}\right)$. Therefore, $a = \text{haversin}(\Delta\phi) + \cos\phi_1 \cos\phi_2 \text{haversin}(\Delta\lambda)$ (Rick). While knowing this formula does not help with finding one's latitude and longitude based on celestial objects, it provides us with the method to find the distance after we have done so. Furthermore, understanding GCDs is essential to understanding the spherical geometry and trigonometry required for celestial navigation, where spherical triangles constructed of GCDs are used.

The formula does have its limitations, namely with its assumption of the spherical shape of the Earth. The Earth is an oblate spheroid with its radius varying from 6356 km 6378 km (Sharp). While this may be unacceptable to an investigation that requires more accuracy in calculations of distances on Earth, that is not the main purpose of this investigation thus this level of inaccuracy is acceptable for the purposes of this investigation. Furthermore, the uncertainties are reduced when calculating long distances. Assuming the Earth is a perfect sphere also allows for simpler visualization of how the sky fits into finding one's latitude and longitude, which is now possible with an understanding of GCDs.

**Calculating coordinates method 1: Solar noon and chronometer**

One of the oldest methods of calculating latitude, ever since we invented instruments to measure the altitude of celestial objects, is to use the altitude of the sun at its highest point and knowledge of the sun's declination. The Earth's rotational axis is tilted 23.45° off its orbital plane and since sun rays hit the Earth parallel to each other, we know that the declination of the sun will vary between -23.45° and 23.45° ("Determining Latitude" [05:55]). Those declination angles correspond to the winter and summer solstices respectively, and the spring and autumn equinoxes occur when the declination angle is 0°. Using the approximate dates for solar declination from UCSB Geography, we can plot a sinusoidal graph to predict the declination angle for any given day.

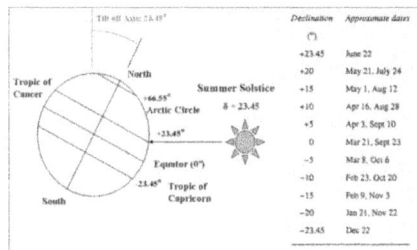

*Figure 8: Solar declination approximate dates ("Insolation")*

Table 1 and Graph 1: Solar declination vs Number of Days since January 1st

| Number of days since January 1st n / days | Solar declination δ / ° |
|---|---|
| 20 | -20.00 |
| 53 | -10.00 |
| 67 | -5.00 |
| 80 | 0.00 |
| 93 | 5.00 |
| 106 | 10.00 |
| 121 | 15.00 |
| 141 | 20.00 |
| 173 | 23.45 |
| 205 | 20.00 |
| 224 | 15.00 |
| 240 | 10.00 |
| 253 | 5.00 |
| 266 | 0.00 |
| 279 | -5.00 |
| 293 | -10.00 |
| 326 | -20.00 |
| 356 | -23.45 |

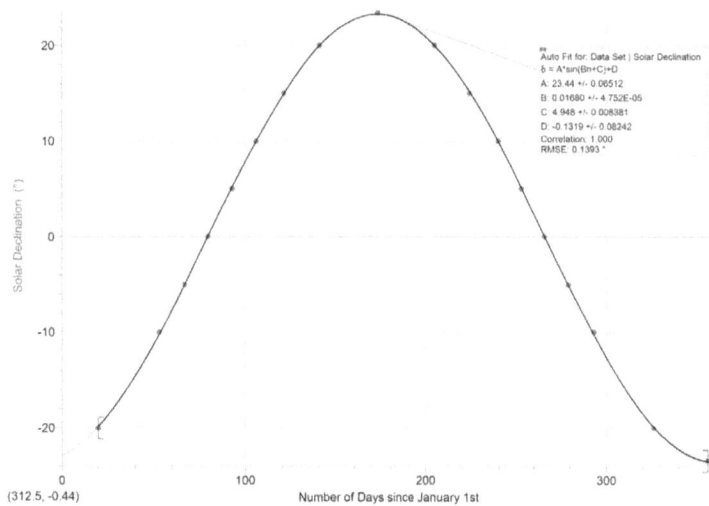

Sinusoidal fit: $\delta = 23.44\sin(0.01680n + 4.948) - 0.1319$

The sinusoidal function from the graph above allows for the prediction of what the solar declination may be for any given day. This model is not perfect, since it assumes that the Earth's orbit is a perfect circle. In reality, Earth's orbit has an eccentricity value of around 0.0167, with 0 being a perfect circle. (Nelson). There have been better models developed to reduce the uncertainties in predicting the solar declination that are far more complex than a sinusoidal model. However, like the uncertainties with the radius of the Earth, these uncertainties are acceptable for the purposes of this investigation and the accuracy of the results are unlikely to be affected to a notable extent.

*Figure 9: CamSextant altitude measurement of the sun at 12:00pm on 28/04/2020*

At solar noon on the 28th of April 2020, I was able to measure the altitude of the sun using a phone app called CamSextant by Omar Reis. Reis' phone app uses the phone's gyroscope sensor to create an artificial horizon instead of an observable horizon a navigator would be able to see at sea which is not as accurate as a traditional sextant. Again, this is another source of uncertainty acceptable for the explorative purposes of this investigation. Since 2020 is a leap year and the measurement was taken in the middle of the day, the Number of Days since January 1st $n$ = 119.5 (the 28th of April is the 119th day of 2020).

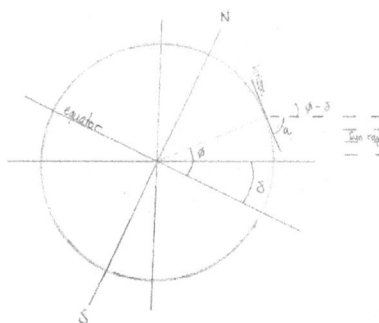

*Figure 10: Solar noon declination and altitude in the northern hemisphere*

From Figure 10, we see that, with $\phi$ as latitude, $\delta$ as solar declination and $\alpha$ as solar altitude:

$$(\phi - \delta) + \alpha = 90$$
$$\phi = 90° - \alpha + \delta$$

108

Substituting in the values of $\alpha$, $\delta$ and $n$ we get:

$$\phi = 90^{\circ} - a + (23.44\sin(0.01680n + 4.948) - 0.1319)$$
$$\phi = 90^{\circ} - 83.43^{\circ} + 23.44\sin(0.01680 \times 119.5 + 4.948) - 0.1319) = 21.038^{\circ}$$

The method for calculating latitude using solar declination and altitude is one of the oldest methods known. It has advantages in its simplicity – all you need is the date, the declination of the sun that day and the altitude of the sun. However, it is much more difficult for a navigator to get a fix on their position since the only measurements they can use are based on the sun – calculations using other celestial bodies cannot be used for comparison. This is why modern celestial navigation often employs multiple methods in tandem with the solar noon method to calculate latitude.

As for longitude, the problem was solved for navigators when the first accurate clock was invented. Navigators could simply set the clock to the time of their location of departure and then check that clock when it is noon wherever they are at sea. Subtracting the time of solar noon at the current location from the 12:00 will yield the time zone difference between the two locations and from there the longitude of the new location can be calculated.

Note that the 'old' methods of calculating latitude and longitude are specific to the situation of solar noon and is limited in its versatility in being applied for other celestial bodies at other times during the day. Calculating longitude by chronometer also requires that one has traveled a given distance and knows their longitude of departure. I am at a fixed position in this investigation therefore this method is not applicable therefore a more versatile method is needed.

**Calculating coordinates method 2: The navigational triangle**

Using a combination of measurements from the equatorial coordinate system and the horizontal coordinate system, a triangle called the navigational triangle or the PZX triangle can be constructed (Davidson). The points P, Z and X represent the Celestial North Pole, the zenith of the observer and the position of the celestial body being observed respectively (Machtelinckx). Connecting these points, the angular distance sides of the spherical triangle can be made.

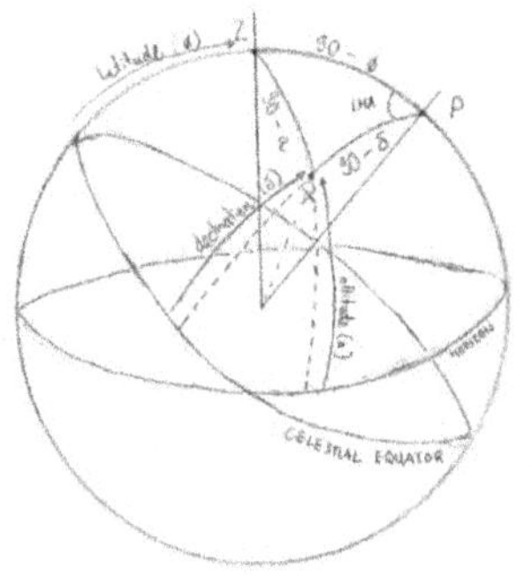


*Figure 11: The Navigational PZX triangle*

The celestial body's declination is measured from the celestial equator to X therefore the angular distance XP is $90 - \delta$. Its altitude is measured from the horizon to the X therefore the angular distance XZ is $90 - a$. The observer's latitude is measured from the equator to Z therefore the angular distance ZP is $90 - \phi$. ∢ZPX is the Local Hour Angle (*LHA*) of the celestial body which is the angle between the observer's meridian and the celestial body's meridian. To solve the PZX triangle and calculate my geographic coordinates, the spherical law of cosines is needed. ("Spherical Law of Cosines").

Consider a spherical triangle ABC with sides a, b and c opposite to each of the angles respectively. By definition, a triangle on a spherical surface is made up of great circle distances and not of lines like a planar triangle. The center of each of the great circle distances $AB$, $BC$ and $AC$ is the center of the sphere O. Note that the side lengths a, b and c are angular distances, corresponding to the central angles $\angle BOC$, $\angle AOC$ and $\angle AOB$ respectively ("Spherical Law of Cosines").

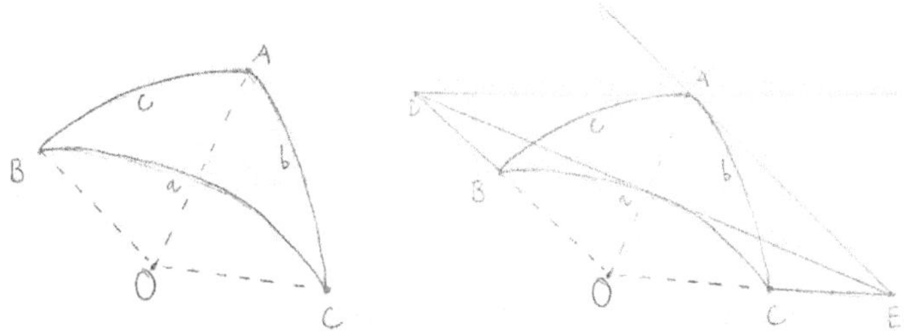


*Figure 12*

Tangent to great circles AB and AC, $\overline{AD}$ and $\overline{AE}$ are drawn respectively.

Therefore $OA \perp \overline{AD}$, $OA \perp \overline{AE}$ (radius tangent theorem).

OB and OC are extended such that they intersect $\overline{AD}$ and $\overline{AE}$ at D and E on separate planes respectively.

Spherical angle $\sphericalangle BAC = \angle DAE = A$ (definition of a spherical angle).

Since spherical triangle sides are angular distances, in planar right triangle AOD, $\angle AOD = \angle AOB = c$

By right triangle trigonometric ratio $\sec\theta = \dfrac{1}{\cos\theta}$ we have:

$$\tan c = \frac{\overline{AD}}{\overline{OA}} \ \therefore \ \overline{AD} = \overline{OA}\tan c$$

$$\cos c = \frac{\overline{OA}}{\overline{OD}} \ \therefore \ \sec c = \frac{\overline{OD}}{\overline{OA}} \ \therefore \ \overline{OD} = \overline{OA}\sec c$$

And similarly, for planar right triangle AOE:

$$\overline{AE} = \overline{OA}\tan b$$
$$\overline{OE} = \overline{OA}\sec b$$

From applying the planar law of cosines to triangle DAE we get:

$$\left(\overline{DE}\right)^2 = \left(\overline{AD}\right)^2 + \left(\overline{AE}\right)^2 - 2\overline{AD}\times\overline{AE}\cos\angle DAE$$

$$\left(\overline{DE}\right)^2 = \left(\overline{OA}\tan c\right)^2 + \left(\overline{OA}\tan b\right)^2 - 2(\overline{OA}\tan c)(\overline{OA}\tan b)\cos A$$

$$\left(\overline{DE}\right)^2 = \left(\overline{OA}\right)^2(\tan^2 c + \tan^2 b - 2\tan b\tan c\cos A)$$

Applying the planar law of cosines to triangle DOE, considering that $\angle DOE = \angle BOC = a$, we get:

$$(\overline{DE})^2 = (\overline{OD})^2 + (\overline{OE})^2 - 2\overline{OD}\times\overline{OE}\cos\angle DOE$$

$$(\overline{DE})^2 = (\overline{OA}\sec c)^2 + (\overline{OA}\sec b)^2 - 2(\overline{OA}\sec c)(\overline{OA}\sec b)\cos a$$

$$(\overline{DE})^2 = (\overline{OA})^2(\sec^2 c + \sec^2 b - 2\sec b\sec c\cos a)$$

Equating the two equations for $\left(\overline{DE}\right)^2$:

$$\tan^2 c + \tan^2 b - 2\tan b\tan c\cos A = \sec^2 c + \sec^2 b - 2\sec b\sec c\cos a$$

Using the Pythagorean trigonometric identity $\sec^2\theta = 1 + \tan^2\theta$:

$$\tan^2 c + \tan^2 b - 2\tan b\tan c\cos A = (1 + \tan^2 c) + (1 + \tan^2 b) - 2\sec b\sec c\cos a$$

$$-2\tan b\tan c\cos A = 2 - 2\sec b\sec c\cos a$$

$$-\tan b\tan c\cos A = 1 - \sec b\sec c\cos a$$

$$-\frac{\sin b}{\cos b}\times\frac{\sin c}{\cos c}\cos A = 1 - \frac{1}{\cos b}\times\frac{1}{\cos c}\cos a$$

Multiplying both sides by $\cos b\cos c$:

111

$$-\sin b \sin c \cos A = \cos b \cos c - \cos a$$
$$\cos a = \cos b \cos c + \sin b \sin c \cos A$$

Substituting the values of the PZX triangle into the spherical law of cosines:

$$\cos(90-a) = \cos(90-\phi) \times \cos(90-\delta) + \sin(90-\phi) \times \sin(90-\delta) \times \cos(LHA)$$

And using the trigonometric identities $\cos(90-\theta) = \sin\theta$ and $\sin(90-\theta) = \cos\theta$:

$$\sin a = \sin\phi \sin\delta + \cos\phi \times \cos\delta \times \cos(LHA)$$

We now have the spherical law of cosines with the side and angles of the PZX triangle substituted in. Now we need a celestial body and measurements of its altitude and declination to use for the calculation of longitude.

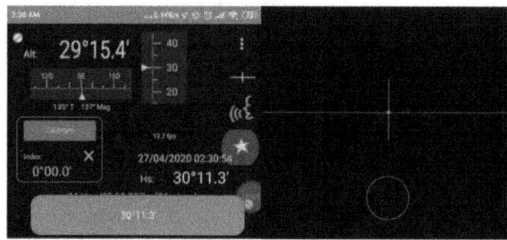

*Figure 13: CamSextant altitude measurement of Jupiter at 2:30AM on 27/04/2020*

On the 27th of April 2020 at 2:30 AM, using the CamSextant app developed by Omar Reis, I sighted Jupiter and measured its angle of altitude ($a$). From there, there are two main ways of solving the navigational triangle: substituting the values for the angles and angular distances for one sight of a celestial body or using an assumed latitude to solve for longitude.

Thus, in a sense, with celestial navigation you often need to know where you were in order to figure out where you are. This may seem like a limitation of the methodology, but in reality, sailors are never completely lost at sea and can work out an estimate for where they are based on the bearing they are travelling on and the speed of the vessel which mostly stays constant. Furthermore, the existence of simpler methods such as latitude by solar noon and longitude by chronometer ensure that today's sailors can always have some idea of where they are.

With the altitude of Jupiter known, we need its declination ($\delta$) and its Greenwich Hour Angle (*GHA*). Both values can be found in a Nautical Almanac, a publication that lists the positions of celestial bodies for navigators to use (Rodegerdts). The Almanac lists the positions in terms of time in UTC (Coordinated Universal Time). A navigator would keep an accurate UTC clock on board however that is not applicable to my hypothetical situation therefore I will have to convert GMT+7 time to UTC, meaning that my sight was at 19:30 on 26/04/2020 UTC. Since the Earth does not orbit Jupiter and the distance from Earth to Jupiter is many times greater than that of the distance to the Sun, the declination ($\delta$) of Jupiter only changes by around 0.1 minutes every hour and was around -20°55.4' at 2am UTC. The *GHA* of Jupiter changes significantly more and when plotted against time we get the following linear graph.

Table 2 and Graph 2: Greenwich Hour Angle vs time for Jupiter on 26/04/2020 UTC

| time t / hours | Greenwich Hour Angle GHA / ° |
|---|---|
| 0 | -84.30 |
| 1 | -69.26 |
| 2 | -54.23 |
| 3 | -39.19 |
| 4 | -24.15 |
| 5 | -9.11 |
| 6 | 5.93 |
| 7 | 20.97 |
| 8 | 36.00 |
| 9 | 51.04 |
| 10 | 66.08 |
| 11 | 81.12 |
| 12 | 96.16 |
| 13 | 111.20 |
| 14 | 126.23 |
| 15 | 141.27 |
| 16 | 156.31 |
| 17 | 171.35 |
| 18 | 186.39 |
| 19 | 201.43 |
| 20 | 216.46 |
| 21 | 231.50 |
| 22 | 246.54 |
| 23 | 261.58 |

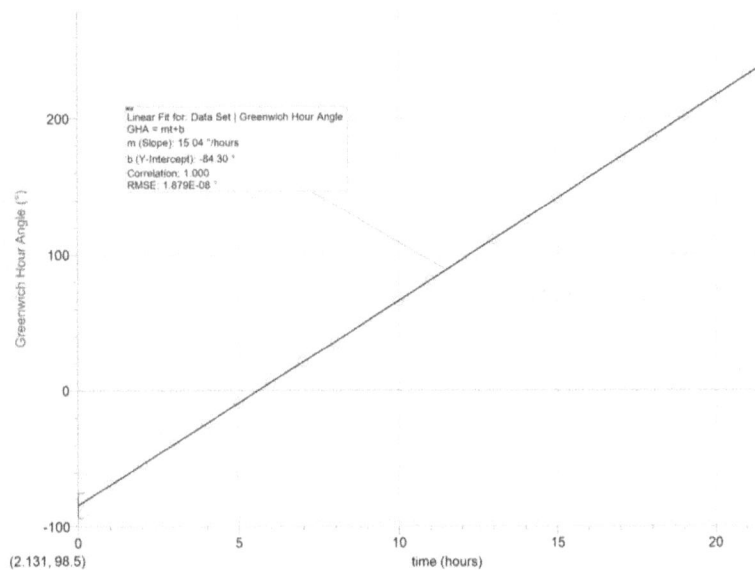

Linear Fit for: Data Set | Greenwich Hour Angle
GHA = mt+b
m (Slope): 15.04 °/hours
b (Y-Intercept): -84.30 °
Correlation: 1.000
RMSE: 1.879E-08 °

Note that GHA is not negative: negative values are present to better illustrate the linear relationship. For example: a GHA of -84.30° (84.30° measured eastward) is a GHA of 275.50° measured westward.

The formula of the linear best fit for GHA vs time is $GHA = 15.04t - 84.30$. The slope is 15.04 which makes sense intuitively since the GHA changes with Earth's rotation therefore in 24 hours Jupiter's GHA should change by around 360° every 24 hours or around 15° every hour. Substituting the time of the sight as 19:30 (19.5 hours) into the equation we get:

$$GHA = 15.04 \times 19.5 - 84.30 = 208.98°$$

113

Rearranging the spherical law of cosines with the sides of the PZX triangle substituted in and calculating for LHA:

$$LHA = \arccos\left(\frac{\sin a - \sin \phi \sin \delta}{\cos \phi \cos \delta}\right)$$

$$LHA = \arccos\left(\frac{\sin(29.257) + \sin(21.038)\sin(-20.923)}{\cos(21.038)\cos(-20.923)}\right) = 44.96°$$

Looking back at Figure 11, we can then see how the LHA and GHA can be used to find the longitude of the observer. The relationship between LHA, GHA and longitude will vary between sights.

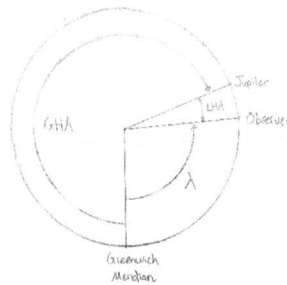

*Figure 14*

Jupiter was observed to be towards the east of the observer which is counterclockwise from the observer in Figure 14. Since Jupiter is east of the observer, we know that the LHA of Jupiter is not part of the GHA (which would be the case if Jupiter was west of the observer). From there, we can deduce:

$$\lambda = 360 - (GHA + LHA)$$
$$\lambda = 360 - (208.98 + 44.96) = 106.06°$$

Therefore, from the latitude calculated from the solar noon method, an altitude measurement of Jupiter and the GHA and LHA of Jupiter taken from the Nautical Almanac, the longitude has been calculated. Alternatively, two sights can be taken, and a system of equations can be set up to solve for latitude and longitude at the same time. It is important to note that navigators will rarely ever settle for a single calculation of their latitude and longitude and will often use a method known as the line of position method to get a 'fix' on their location based on multiple sights. The mathematics behind the line of position method and the working out of the probability of where the observer is likely to be relative to the lines of position drawn is beyond the scope of this investigation however for further reading please take a look at Onboard Intelligence linked on the Works Cited page at the bottom of the investigation.

Even with the best equipment available resulting in significantly reduced random and systematic errors, real navigators still need to use such methods to be as accurate as possible in their calculations. Using the CamSextant app to make my altitude measurements means that the inaccuracies in my calculation of longitude would likely not be acceptable to a real navigator. Nonetheless, these practical measurements allow for the hypothetical nature of this investigation to be grounded in reality to a certain extent.

**Initial Bearing**

In this hypothetical situation, I now know how to find my coordinates and calculate the shortest distance between my point of departure and my destination. The final thing I will need to know if I am to embark on such a journey is the direction I need to travel. Instead of using spherical trigonometry, this problem can be tackled using vector analysis (Rick).

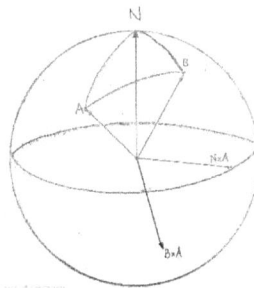

*Figure 15: Three-dimensional vector diagram for two points and North Pole on a sphere*

Again, we use a unit sphere to derive the formula first. We have three unit vectors that each start in the center of the sphere that each point in the direction of a point on the sphere: N in the direction of the North Pole, A in the direction of the starting point, and B in the direction of the destination point. The starting point A has spherical coordinates $(\phi_1, 0)$ and point B has coordinates $(\phi_2, \Delta\lambda)$.

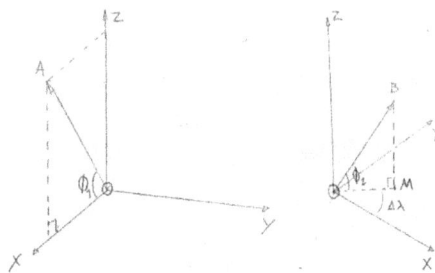

*Figure 16*

115

Point N is parallel to the Z-axis therefore $\overrightarrow{ON} = \begin{pmatrix} 0 \\ 0 \\ 1 \end{pmatrix}$

Point A has longitude 0 therefore we can place it on the X-axis with zero Y displacement. From there we get

$OX = AO\cos\phi_1$ and since this is a unit sphere, AO = 1 thus $OX = \cos\phi_1$

$AX = AO\sin\phi_1$ and since this is a unit sphere, AO = 1 thus $AX = \sin\phi_1$

OX is the displacement of $\overrightarrow{OA}$ on the X-axis and AX is the Z component of $\overrightarrow{OA}$ therefore $\overrightarrow{OA} = \begin{pmatrix} \cos\phi_1 \\ 0 \\ \sin\phi_1 \end{pmatrix}$

For vector $\overrightarrow{OB}$, $\angle XOM = \Delta\lambda$ since the X-axis corresponds to the longitude 0 of point A. Since OB is a radius of the unit sphere, we have $\sin\phi_2 = \dfrac{BM}{OB} \therefore BM = \sin\phi_2$ and $\cos\phi_2 = \dfrac{OM}{OB} \therefore OM = \cos\phi_2$

In triangle XOM we have $OX = \cos\Delta\lambda\cos\phi_2$

In triangle YOM we have $OY = \cos(90-\Delta\lambda)\cos\phi_2 = \sin\Delta\lambda\cos\phi_2$

BM, OX and OY are the X, Y and Z components of $\overrightarrow{OB}$ respectively therefore: $\overrightarrow{OB} = \begin{pmatrix} \cos\Delta\lambda\cos\phi_2 \\ \sin\Delta\lambda\cos\phi_2 \\ \sin\phi_2 \end{pmatrix}$

Taking the vector products NxA and BxA we get:

$$N\times A = \begin{pmatrix} 0 \\ 0 \\ 1 \end{pmatrix} \times \begin{pmatrix} \cos\phi_1 \\ 0 \\ \sin\phi_1 \end{pmatrix} = \begin{pmatrix} \sin\phi_1\times 0 - 0\times 1 \\ \cos\phi_1\times 1 - \sin\phi_1\times 0 \\ 0\times 0 - \cos\phi_1\times 0 \end{pmatrix} = \begin{pmatrix} 0 \\ \cos\phi_1 \\ 0 \end{pmatrix} \text{ and}$$

$$B\times A = \begin{pmatrix} \cos\Delta\lambda\cos\phi_2 \\ \sin\Delta\lambda\cos\phi_2 \\ \sin\phi_2 \end{pmatrix} \times \begin{pmatrix} \cos\phi_1 \\ 0 \\ \sin\phi_1 \end{pmatrix} = \begin{pmatrix} \sin\Delta\lambda\cos\phi_2\times\sin\phi_1 - \sin\phi_2\times 0 \\ \sin\phi_2\times\cos\phi_1 - \cos\Delta\lambda\cos\phi_2\times\sin\phi_1 \\ \cos\Delta\lambda\cos\phi_2\times 0 - \sin\Delta\lambda\cos\phi_2\times\cos\phi_1 \end{pmatrix} = \begin{pmatrix} \sin\Delta\lambda\cos\phi_2\sin\phi_1 \\ \sin\phi_2\cos\phi_1 - \cos\Delta\lambda\cos\phi_2\sin\phi_1 \\ -\sin\Delta\lambda\cos\phi_2\cos\phi_1 \end{pmatrix}$$

Since NxA and BxA are perpendicular to the great circles with arcs NA and NB respectively, the angle between NxA and BxA is equal to the angle we are looking for, the bearing of B from A or $\angle ANB$.

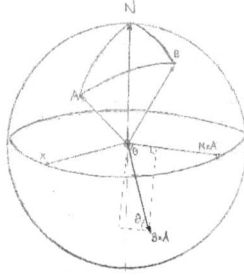

*Figure 17*

A has longitude 0 and zero Y-displacement therefore NxA, which is perpendicular to A, is parallel to the Y-axis. Thus, we have a right triangle with side lengths of the Y component of BxA adjacent to angle $\theta$ and the component of BxA in the XZ plane opposite to $\theta$. From using the Pythagorean theorem to find the component of BxA in the XZ plane and taking the tangent of $\theta$ using the known opposite and adjacent sides we get:

$$\tan\theta = \frac{\sqrt{(\sin\Delta\lambda\cos\phi_2\sin\phi_1)^2 + (-\sin\Delta\lambda\cos\phi_2\cos\phi_1)^2}}{\sin\phi_2\cos\phi_1 - \cos\Delta\lambda\cos\phi_2\sin\phi_1}$$

$$\tan\theta = \frac{\sqrt{(\sin\Delta\lambda)^2(\cos\phi_2)^2}\sqrt{((\sin\phi_1)^2 + (\cos\phi_1)^2)}}{\sin\phi_2\cos\phi_1 - \cos\Delta\lambda\cos\phi_2\sin\phi_1}$$

And since $((\sin\phi_1)^2 + (\cos\phi_1)^2) = 1$ :

$$\tan\theta = \frac{(\sin\Delta\lambda)(\cos\phi_2)}{\sin\phi_2\cos\phi_1 - \cos\Delta\lambda\cos\phi_2\sin\phi_1} \quad \therefore \theta = \arctan(\frac{(\sin\Delta\lambda)(\cos\phi_2)}{\sin\phi_2\cos\phi_1 - \cos\Delta\lambda\cos\phi_2\sin\phi_1})$$

Which gives us the initial bearing angle of point B from A given the latitude and difference in longitude of the two points.

**Conclusion**

In the introduction of this investigation, I stated that the two things I will be calculating from my coordinates found through celestial navigation would be the distance from my location to Ho Chi Minh City and the direction I need to travel in. Shown below is the arbitrary point in Ho Chi Minh City from Google Maps that will be used to compare the calculations made with Google Maps' coordinates with those made with my calculated coordinates.

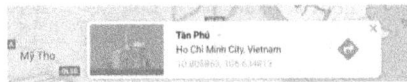

*Figure 18: Google Maps coordinates of an arbitrary point in Ho Chi Minh City (Google)*

Substituting the calculated latitude and longitude of my position and the latitude and longitude of Ho Chi Minh City according to Google Maps into the Haversine formula from earlier in the investigation we get:

$$a = \sin^2\left(\frac{10.80563 - 21.038}{2}\right) + \cos 10.80563 \cos 21.038 \sin^2\left(\frac{106.634813 - 106.06}{2}\right) = 0.00788914$$

$$\theta = 2\arctan\left(\frac{\sqrt{a}}{\sqrt{1-a}}\right) = 0.177876$$

$$d = R \times \theta = 6371 \times 0.177876 = 1133.24 \text{ km}$$

Substituting the calculated latitude and longitude of my position and the latitude and longitude of Ho Chi Minh City according to Google Maps into the initial bearing formula we get:

$$\theta = \arctan\left(\frac{(\sin\Delta\lambda)(\cos\phi_2)}{\sin\phi_2\cos\phi_1 - \cos\Delta\lambda\cos\phi_2\sin\phi_1}\right)$$

$$\theta = \arctan\left(\frac{(\sin(106.634813 - 106.06))(\cos 10.80563)}{\sin 10.80563\cos 21.038 - \cos(106.634813 - 106.06)\cos 10.80563\sin 21.038}\right) = -3.17545$$

The result is in radians therefore converted to degrees we get $\left(-3.17545 \times \frac{180}{\pi}\right) = -181.94°$

Which is 181.94° measured counterclockwise from north therefore the bearing is 178.06°.

According to Google Maps, the distance from my actual position to the coordinates of Ho Chi Minh City is 1138.69 km. Calculating the percent error of my calculation based on my coordinates obtained from measurements of the Sun and Jupiter we get:

$$\frac{|1133.24 - 1138.69|}{1138.69} \times 100\% = 0.48\%$$

*Figure 19: Distance from Hanoi to HCMC (Google)*

Overall, a percent error of 0.48% is acceptable for this investigation and successfully shows that calculations revolving around celestial navigation are quite accurate when compared to what GPS navigation tells us.

118

This investigation set out to explore the mathematics behind celestial navigation and compare the results of those calculations to those of GPS navigation. It essentially set out to answer three questions anyone who is trying to get somewhere would ask: 'where is my starting point, how far is it to where I need to go and what direction do I need to travel in?' Through using the spherical law of cosines, the Haversine formula and vector analysis, the investigation was able to successfully touch on all three of those questions. I now have an understanding and appreciation for celestial navigation, one of the most widespread practical applications of celestial coordinates we see today. Even after building a horizontal coordinate system telescope, the relationship between the celestial coordinates and coordinates on Earth was not clear to me and only became clear through a mathematical investigation. Furthermore, we also see how spherical geometry can describe the world around us in ways that planar geometry is unable to.

Obviously, there are many flaws with the methodology of this investigation and its somewhat limited scope. First, although it has been mentioned throughout the investigation, the lack of access to proper resources hinders the investigation. Although Omar Reis' CamSextant app is a decent alternative to a real sextant, it is significantly more inaccurate and is somewhat counterintuitive in an investigation centered around navigation without the use of digital technology. Furthermore, as stated in a comment under the navigational PZX triangle method, the scope of the investigation does not include methods used by navigators to reduce those uncertainties. For example, although the core mathematics such as the spherical law of cosines was touched on, the intercept method, which reduces many uncertainties, is still an important step that navigators always take today. There are also many other calculations that navigators factor in, such as refraction of light from the celestial body and the height of the ship. The calculations for distance, bearing, and latitude could also be even be more accurate if the true shape of the Earth and its orbit was factored in.

Despite those limitations, the purpose of this investigation was not to construct a fool-proof method of for everything one needs to navigate based on celestial objects. It was to investigate the mathematics behind a non-digital alternative to GPS navigation and compare how it holds up. All of the aforementioned extensions to reduce the uncertainties of the calculations can be found in the Works Cited list.

I have definitely not learned everything there is to learn about celestial navigation, but I now understand the core mathematical concepts behind it. Although, if I was somehow lost at sea equipped with a sextant, a compass, an almanac, a map, a clock, and a calculator (all things most ships would have on deck when travelling long distances), I can now confidently say that I would probably be able to find my way back to land and know how long the journey would take.

# 4. THE LIQUIDITY OF CATS

Author: Sarah Florentine Sinartio
Moderated Mark: 17/20
Level: Math AA SL

I.    Introduction

Whether you have cats yourself, are an avid cat lover, or are just simply on Instagram a lot, there has been a recent popularity in one of the odder characteristics of a cat, this being its "liquidity." As explained by Marc-Antoine Fardin (Ig Nobel Prize Winner in physics), his exploration of the rheology of cats explained how cats are able to become liquid when given the time to. As liquids can be defined as "a material that adapts its shape to fit a container" cats have been repeatedly seen being able to fit in boxes that are visually smaller than their volume, as seen in Figure 1 and Figure 2 (Marc-Antoine Fardin).

Figure 1: Large cat, small box
(Lebatihem)

Figure 2: Big Cat Shoe Box (bnilsen)

As someone who has 10 cats herself I have seen this phenomenon first-hand multiple times as when it rains my cats like to sleep in cardboard boxes. Sometimes when I forget to place enough cardboard boxes out I see 10 of them sharing 2 medium-sized boxes, with 5 in each. Due to this, I began to question its limits; is there a definite limit to the volume a cat can fit in? My aim is to investigate the minimum volume of a cat and to use this minimum volume to determine whether or not cats are "liquid" or not by observing whether or not they are able to occupy a box 25% and 50% smaller than their volume.

II.  Investigation

A.  Method of Approximation

I began to search the different ways in which I can measure the volume of a cat, firstly I just decided to find boxes that would best fit each cat. It is important to note that throughout this investigation, I will be using the three same cats: Mama Pasta, Penne, and Pesto out of my 10 cats. Each of these cats differ in their weight and age with Mama Pasta being the oldest (around 2 years old) therefore the heaviest in weight (5.1 kg), Penne being the second oldest (9 months old) and weighing 3.1 kg, and Pesto is the youngest (5 months old) weighing at 1.9 kg. Below in Table 1, I scavenged for boxes around my house that each cat could fit in and noted down the dimensions of the box to find the volume.

*Table 1: Method of Approximation*                     *Figure 3: Pesto in the Box*

| Cat Name | Box Dimensions/cm | Volume/ cm$^3$ | Weight/ Kg |
|---|---|---|---|
| Mama Pasta | 24.5 x 19.0 x 20.4 | 9496.2 | 5.1 |
| Penne | 21.2 x 18.6 x 18.1 | 7137.2 | 3.1 |
| Pesto | 14.0 x 16.0 x 14.2 | 3180.8 | 1.9 |

To evaluate the method of approximation, as we can see in the photo above (Figure 3) the cats would not take up all the provided space in the boxes. This would mean that the volumes found are going to be more than the actual volume of the cat. In addition to this, in terms of the methodology itself, depending on the position the cat has decided to take inside the box, it can affect the amount of extra space leftover. For example, for Pesto, he repeatedly used the same position every time he entered the box and would not allow me to change his position to ensure that the maximum space is being occupied by him. Therefore, there is an issue of preference with the position the cats take. Also another limitation is that the dimensions of the boxes are limited to the boxes that I have at home and therefore might not be the smallest box that the cats can fit but in fact the smallest box that the cat can fit out of the boxes that I have at home.

### B. Water Displacement Method

The next most common method that I knew of to measure volume was the water displacement method. Which I attempted to do with the cats as seen in Table 2 but was only able to do for Pesto. With this method, I filled a container, with a known volume, with water. I then placed this container inside a larger container. I would then proceed to attempt to fully submerge the cat in the container with water and measure the volume of the displaced water (which would be in the larger container).

*Table 2: Water Displacement Method*

| Cat Name | Initial Volume of Water /cm$^3$ | Volume of Displaced Water (Volume of Cat) /cm$^3$ |
|----------|-------------------------------|--------------------------------------------------|
| Pesto | 13440 | 2300 |

There were two main issues in this method. The first one being that in order to properly measure the volume of the cat using this method I would have to fully submerge the cat which as we can see in Figure 4, is impossible.

Figure 4: Measuring Pesto's Volume using Water Displacement Method

A vast majority of cats despise water because of this it makes it difficult to submerge them. Pesto also began to bash his paws frantically and this caused excess water to exit the original bucket. This will affect the accuracy of the volume of the cat. Not only that, but because of Pesto's hatred of water, he became very scared and began to scratch things like my foot (as seen in Figure 5). I decided that this method was not appropriate and very difficult to do so I decided to not implement this method on the other cats but instead to search for better methods.

Figure 5: Disadvantages of Water Displacement Method

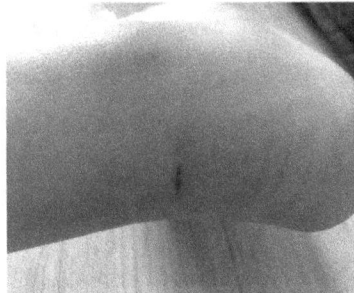

C. Washer/disc Method

After a while, I came across a prezi presentation where they used GeoGebra to determine the volume of a Somoan boat (William). They did so by creating different curves that outline the shape of the boat which they then found the area and the volume using the washer/disc method. The washer/disc method is a common way to determine the volume of a solid of a revolution. Since both the Samoan boat and my cats are irregularly shaped, I decided that this would be my next course of action.

Before I began I decided to note down the measurements of each of the three cats (measured using tape measure), which can be seen in Appendix A. This is to ensure that the scale from picture to graph is appropriate. For example, in order to determine the area and volume of each of the cat's faces, I stretched and widened the image (of the cat's face) to ensure that the face length of the cat on the graph is the same as recorded in Appendix A. Therefore 1 unit in GeoGebra equates to 1 cm.

After measuring the dimensions of the cats, I took a picture of the following parts of the cat: face, body, leg, paws, tail, neck. I decided not to include the ears of the cats because they are bendable. With these pictures I inserted each to GeoGebra and aligned it based on the measurements in Appendix A symmetrically with the line of symmetry being y = 0. Then I traced the body part (from y ≥ 0) by plotting points, which will be used to find the best fit curve. To walk through the manual calculations of the volume of each component, we will take Pesto's head as an example. Since we are using the washer/disc method, we must obtain a solid of revolution. In order to obtain this, we must start with a function, which we can define as y = f(x). y = f(x) can be found using the 'FitPoly' code on GeoGebra. FitPoly generates a polynomial equation, which will be our f(x), based off of a set of points plotted. An example of the FitPoly code can be seen in the Figure (13) below:

*Figure 6: FitPoly Code Example*

$$f(x) = FitPoly(\{C, D, E, F, G, H, I, J, K, L, M, N, O, P, Q, R\}, n)$$

In this example code {C, D, E, F, G, H, I, J, K, L, M, N, O, P, Q, R} represents each of the points plotted on the graph and n represents the degree of the polynomial, which can be chosen specifically for each body part. Since we already have the points plotted from the previous method, we can automatically input the FitPoly to obtain our equation.

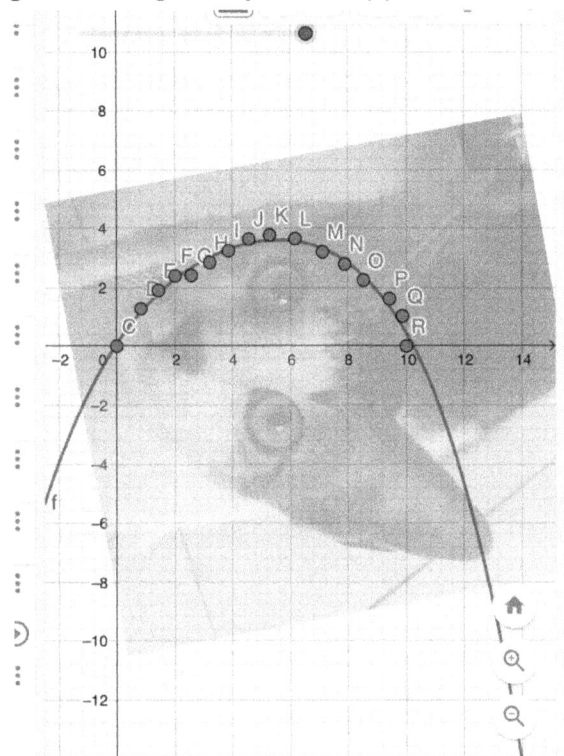

Figure 7: Using FitPoly to Find f(x) for Pesto's Head

Using the slider feature on GeoGebra, I chose n = 4.1. This is because the curve with the degree of polymerization at 4.1 (rounded to 4) is the best fit curve. Meaning that many of the points plotted (through tracing Pesto's head) fall near or on the curve. The function, f(x), obtained is the following:

$$f(x) = -0.0011972x^4 + 0.020253x^3 - 0.22694x^2 + 1.47x + 0.081572$$

Proceeding this, we need to place parameters on the function. This is to ensure that when we rotate the curve on the x-axis, it will rotate based on Pesto's head. This step is to help visualize the solid of revolution but it does not contribute to the calculation of the volume. This step is achieved using the conditional statement:

$$g(x) = If(x(C) \leq x \leq x(R), f(x)$$

$$C = first\ plot\ point\ on\ f(x)$$
$$R = last\ plot\ point\ on\ f(x)$$

For each different body part, and hence for each different f(x), the 'R' in this conditional statement can be replaced by any letter, depending on the number of plot points on y = f(x). For our current example of Pesto's head the code is as follows:

*Figure 8: Conditional Statement Code for Pesto's Head*

$g(x) = If(x(C) \leq x \leq x(R), f(x))$

$\rightarrow$  $-0.0011972\ x^{4.0000} + 0.020253\ x^{3.0000} - 0.22694\ x^{2.0000} + 1.4764\ x + 0.081572,$  $(0.0000 \leq x \leq 10.000)$

We can now rotate g(x), which will allow us to see the surface of the solid of revolution on the 3D Graphics viewing option in GeoGebra. We will rotate g(x) using two statements of codes, in the same order presented.

$$rotate(g, q, xAxis)$$

and

$$surface(g, q, xAxis)$$

This will generate a slider for q. To rotate g(x), q = 359 will be selected. This will produce the following visualization (Figure 9) to the left of the original graph containing Pesto's picture.

*Figure 9: Solid of Revolution for Pesto's Head*

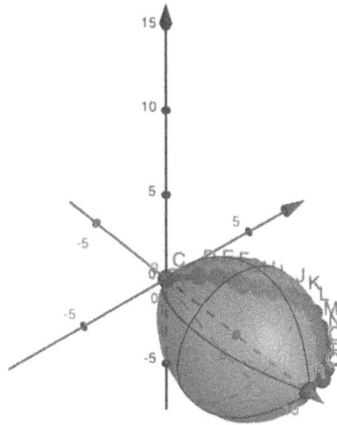

Now that we have the visual representation, we can now proceed with the calculations. This requires us to return to the first stages, to the equation of y = f(x), which we have defined to be the following:

$$f(x) = -0.0011972x^4 + 0.020253x^3 - 0.22694x^2 + 1.47x + 0.081572$$

The needed formula to calculate the volume of this solid of revolution can be derived from the Riemann sum with the limit of $n \to \infty$. Which can be seen below, where V(S) is the volume of the solid of revolution:

$$V(S) = \lim_{n \to \infty} \sum_{i=1}^{n} A(x_i^*)\Delta x$$

From this formula we can further derive the following formula to find the volume of the solid within our pre-determined boundaries of $0 \le x \le 10$:

$$\int_a^b A(x)dx$$
$$A(x) = area = \pi r^2$$

Therefore we can substitute f(x) into r and the boundaries will be a = 0 and b =10 (as measured beforehand in Appendix A) into the following equation:

$$V(S) = \pi \int_a^b (f(x))^2 dx$$

Which will become the following:

$$V(S) = \pi \int_0^{10} (-0.0011972x^4 + 0.020253x^3 - 0.22694x^2 + 1.47x + 0.081572)^2 \cdot dx$$

The equation above can be solved, first by expanding the equation:

$$= \pi \int_0^{10} (0.00000143329x^8 - 0.00004x^7 + 0.00095x^6 - 0.01271x^5 + 0.11085x^4 - 0.66389x^3 + 2.12387x^2 + 0.23982168x + 0.00665) \cdot dx$$

Then, using the sum formula which is the following:

$$f(x) \pm g(x) \cdot dx$$

Which would look like the following considering the equation V(S):

$$V(S) = \pi (\int_0^{10} f(x) \cdot dx \pm \int_0^{10} g(x) \cdot dx)$$

We can finally solve for V(S) by applying the sum rule the equation would look as the following:

$$= \pi (\int_0^{10} 0.00000143329x^8 \cdot dx - \int_0^{10} 0.00004x^7 \cdot dx + \int_0^{10} 0.00095x^6 \cdot dx - \int_0^{10} 0.01271x^5 \cdot dx + \int_0^{10} 0.11085x^4 \cdot dx - \int_0^{10} 0.66389x^{3dx} + \int_0^{10} 2.12387x^2 \cdot dx + \int_0^{10} 0.23982168 \cdot dx + \int_0^{10} 0.00665 \cdot dx)$$

Proceeding this, we will first solve each definite integral. The first step in doing so is to factor out the constant, which for the first definite integral ($\int_0^{10} 0.00000143329x^8 \cdot dx$) is 0.00000143329:

$$\int_0^{10} 0.00000143329x^8 \cdot dx = 0.00000143329 \int_0^{10} x^8 \cdot dx$$

a

130

Subsequently, we can now integrate $x^8$. Which would use the following formula:

$$\int_a^b x^c = [\frac{x^{c+1}}{c+1}]_0^{10}$$

Therefore integrating $x^8$ would yield the following equation, with c = 8:

$$= 0.00000143329 \, [\frac{x^{8+1}}{8+1}]_0^{10}$$

Then we will apply the fundamental theorem of calculus (FTC), which would look like the following:

$$= 0.00000143329 \, [0.00000143329(10)^9 - 0.00000143329(0)^9]$$

Equating the equation above will yield the answer for the integral of $\int_0^{10} 0.00000143329x^8 \cdot dx$. Which is the following:

$$= 159.254$$

These steps can be used to solve the rest of the definite integrals, which is condensed below:

$-\int_0^{10} 0.00004x^7 \cdot dx = -0.00004 \int_0^{10} x^7 \cdot dx = -0.00004[\frac{x^{7+1}}{7+1}]_0^{10} = -0.00004[0.125x^8]_0^{10} = -0.00004 \cdot 12500000 = -500.000$

$\int_0^{10} 0.00095x^6 \cdot dx = 0.00095 \int_0^{10} x^6 \cdot dx = 0.00095[\frac{x^{6+1}}{6+1}]_0^{10} = 0.00095[0.14285x^7]_0^{10} = 0.00095 \cdot 1428571.42857 = 1357.143$

$-\int_0^{10} 0.01271x^5 \cdot dx = -0.01271 \int_0^{10} x^5 \cdot dx = -0.01271[\frac{x^{5+1}}{5+1}]_0^{10} = -0.01271[0.16666x^6]_0^{10} = -0.01271 \cdot 166666.66666 = -2118.333$

$\int_0^{10} 0.11085x^4 \cdot dx = 0.11085 \int_0^{10} x^4 \cdot dx = 0.11085[\frac{x^{4+1}}{4+1}]_0^{10} = 0.11085[0.2x^5]_0^{10} = 0.11085 \cdot 20000 = 2217.000$

$-\int_0^{10} 0.66389x^3 \cdot dx = -0.66389 \int_0^{10} x^3 \cdot dx = -0.66389[\frac{x^{3+1}}{3+1}]_0^{10} = -0.66389[0.25x^4]_0^{10} = 0.66389 \cdot 2500 = -1659.725$

$\int_0^{10} 2.12387x^2 \cdot dx = 2.12387 \int_0^{10} x^2 \cdot dx = 2.12387[\frac{x^{2+1}}{2+1}]_0^{10} = 2.12387[0.33333x^3]_0^{10} = 2.12387 \cdot 333.33333 = 707.957$

$-\int_0^{10} 0.23982168x \cdot dx = -0.23982168 \int_0^{10} x \cdot dx = -0.23982168[\frac{x^{1+1}}{1+1}]_0^{10} = -0.23982168[0.5x^2]_0^{10} = -0.23982168 \cdot 50 = -11.991$

$\int_0^{10} 0.00665 \cdot dx = [0.00665x]_0^{10} = 0.0667$

After solving all of the individual definite integrals, they can now be summed up and multiplied by π:

$$= \pi(159.254 - 500.000 + 1357.143 - 2118.333 + 2217.000 - 1659.725 + 707.957 - 11.991 + 0.067)$$

Which will bring us to Pesto's head's volume at,

$$= 475.549 \ cm^3$$

In Appendix B, C, and D, the same calculations (condensed) can be seen for each of the body parts for Pesto, Penne, and Mama Pasta (respectively) and can also be seen in Table 4. In order to verify the manual calculations of the volumes of each cat, I decided to use Ti-Nspire to re-calculate the volumes. I did so by inputting the following into my Ti-Nspire:

| Figure 10: First step is to press Menu → Calculus → Numerical Integral | Figure 11: Inputting the boundary and equation, to calculate Pesto's head |
|---|---|

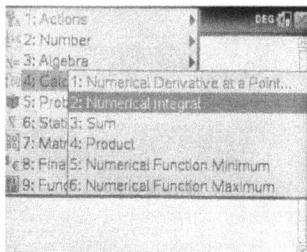

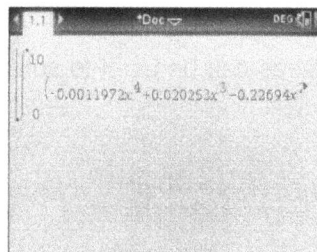

Using the Ti-Nspire, the following volumes were found:

*Table 3: Cats Volume using Ti-Nspire*

| |
|---|
| Pesto<br>= 550.888+4997.944+77.7265+4(39.848)+4(5.51475)+1801.57<br>= 7609.5795 cm³ |
| Penne<br>= 1756.97+5003.642611+256.864+4(253.506)+4(15.6784)+11.3076 |

| = 8105.52181 |
| --- |
| Mama Pasta<br>= 841.215+3451.02+322.988+4(110.03)+4(10.331)+60.37<br>= 5157.037 |

The volumes calculated using the Ti-Nspire were not exactly the same as the manual calculations (refer to Table 4). This is due to the precision of the manual calculations. For some of the manual calculations, the number of significant figures was only 1 whilst the Ti-Nspire calculated the volume using the exact numbers. As a result, I will be using the volume calculated using the Ti-Nspire for the rest of the investigation.

*Table 4: Volume Comparison*

| Cat | Weight / kg | Volume / cm$^3$ | | |
| --- | --- | --- | --- | --- |
| | | Manual | Ti-Nspire | Absolute Difference |
| Pesto | 1.9 | 5586.57244 | 7609.5795 | 2023.01 |
| Penne | 3.1 | 9596.05548 | 8105.52181 | 1490.53 |
| Mama Pasta | 5.1 | 5478.03556 | 5157.037 | 320.999 |

D. Investigating Liquidity

In order to observe the liquidity of a cat, I will be creating boxes with volumes smaller than the calculated volumes of the cat. There will be 2 boxes for each cat, one that is 25% smaller and one that is 50% smaller than the calculated volume. I will then create a box with the dimensions that produce the volume approximately and let the cats attempt to enter the box (with no time limit). There can only be two outcomes, the first one being that the cats do fit in the box (indicated by "Fits in box") or the cats do not (Indicated by "Does not fit in box"). An example (Pesto, 25% smaller) of how the volumes of the boxes will be calculated is shown. Because Pesto's calculated volume is 7609.5795, to find 25% 0.25 must be multiplied by 7609.5795. This will yield:

$$7609.5795 \cdot 0.25 = 1902.394875$$

Then 1902.394875 will be subtracted from the original volume to produce the volume of the box.

$$7609.5795 - 1902.394875 = 5707.184625$$

From this, the dimensions of the box (which will be a cuboid) is the integer of the cube root of the volume. Like the following:

$$\sqrt[3]{5707.184625} = 17.87 \approx 18 \; cm$$

The rest of the volumes of the boxes and the dimensions that will be used can be seen in Table 5 as well as the results.

Table 5: Volumes and Dimensions of Boxes

| Cat Name | Volume / cm³ | |
| --- | --- | --- |
| | 25% Smaller | 50% Smaller |
| Pesto | 5707.184625 | 3804.79 |
| | Fits into box 18 x 18 x 18 | Fits into box 16 x 16 x 16 |
| Penne | 6079.1413575 | 4052.76 |
| | Fits into box 18 x 18 x 18 | Fits into box 16 x 16 x 16 |
| Mama Pasta | 3867.78 | 2578.52 |
| | Fits into box 16 x 16 x 16 | Fits into box 14 x 14 x 14 |

As per the results in Table 5, it was observed that all the cats were able to fit inside the box. Therefore, it can be deduced that cats are able to fit any box hence supporting their liquid property. However, it is important to note that for Mama Pasta it took her quite some time to get situated in the box. Some limitations that I would like to highlight of this method is that there were parts of the cat that did not fit inside the box (e.g. the head). This is all dependent on how the cat chooses to vacate the box. However, I decided that if a majority of their body is within the box then it would be classified as "Fits into box". Therefore, this phenomena and the fact that the calculated dimensions for the boxes were rounded up/down contributes to the fact that the test does not accurately represent the cats' ability to fit in a box of exactly 25% or 50% smaller than their volume.

III.   Discussion and Conclusion

To conclude, my investigation supports the notion that cats have liquid attributes, allowing them to fit in a box of a volume 25% and 50% smaller than their own. It is also found that the approximation and water displacement methods are not appropriate in finding the volume of a cat (or any living being) due to its inaccuracy. With the approximation method the cats did not occupy the whole space of the box and the water displacement method needed the cats to be fully submerged in water. Moreover, the washer/disc method is the most appropriate and accurate method, out of the 3 methods, to calculate the volume of a cat.

In further investigations, it would be better to have a closed box (with holes to allow for respiration). As an extension it would be interesting to see how different shapes affect this liquid property of cats as another one of my observations is that my cats also like to sleep in circular objects. As in my garage there is a hose which is always kept in a circular form (creating a short cylindrical shape) and the cats also always fight for this spot to take a nap in. This is also supported by the fact that there are cat products for semi-circular beds that can be attached to windows. Therefore, the question remains whether or not this property is enhanced or limited from different shaped boxes (circles, triangles, trapezoids, etc.) As Mama Pasta had the lowest volume out of the three cats, although she was the oldest and heaviest, further investigations can explore the relationship between age and volume. In addition to, as previously stated there are limitations in validating the liquidity of the cats as there would be certain body parts that did not fit inside the box. Henceforth, it would be best to create boxes that can close and ensure that this does not happen.

Overall, I never thought that I would be able to relate my love for animals to maths however this investigation has proven me otherwise. It has been a very eye-opening journey into the real-life applications of mathematics, especially calculus and solids of a revolution. I have also recently read an article on the application of solids of a revolution in augmented reality to enhance human to computer interaction (Salinas and González-Mendívil) which allow more realistic animation. Perhaps, my investigation into the volumes of cats may develop further to assist in creating more realistic animations of cats in augmented reality.

In loving memory of Oglio, Bolognese, and Macaroni who have sadly passed during this investigation.

# 5. MODELING PACE AND HEART RATE

Author Parnian Asgari
Moderated Mark: 20/20
Level: Math AA SL

## *Introduction and Rationale:*

There were two things that I did everyday throughout my first year of DP, studying Biology and training for a half-marathon and whenever I did the second activity, I wondered how I could improve my performance and set a better personal record (PR). When searching for an answer I came across heart rate training. In this type of training, you use your heart rate (HR) - in beats per minute (BPM)- as an indicator of intensity, then based on that you create the right workout for your goal. The benefit of this system over measuring your pace is that you will not push yourself too hard and you will not damage your skeletal and muscular systems. At this point my interest in Biology interfered and it was then that the idea of finding a relationship between the pace – the old school method of measuring intensity- and the heart rate, the more recent method of measuring intensity.

The **aim** of this paper is to find a relationship between pace shown in the form of (minutes: seconds) and heart rate measured in beats per minute. There are two reasons for which I have decided to collect first hand data, the first reason is that the complete running data of multiple days of no athlete is found online, and the second reason is that there are many factors affecting both the speed and heart rate and as the person conducting this study, I am the only one who can control them and collect the data in the desired conditions.

Based on my experience as one's pace gets faster, the heart rate should get higher, because as the speed gets higher the heart beats faster, thus my **hypothesis** is that with a faster pace heart rate increases; but whether this is a polynomial, linear, exponential or another type of relation was not obvious to me before finding a model.

## Data collections and results:

The apparatus used to collect the data was a Garmin forerunner 945 multisport watch; this device uses GPS and information such as cadence, 3D speed, 3D distance, etc. to provide the user with the statistics of sports such as running, swimming, body building and more. The data was collected in 1-minute intervals in a period of three weeks and it was then saved into the Garmin connect app. This application provided me with different information such as the pace, heart rate, weather condition, stress levels, recovery rates after each run, $V_{o2}$ max and more. These information were used both for the analysis and for when I wanted to control some variables.

As mentioned earlier there are a number of variables which affect heart rate and pace, below I will state each, explain how they affect the heart rate and the pace and mention how I will control it.

*Sleep:* according to Phyllis Zee, MD, PhD, professor of neurology and director of the Sleep Disorders Program at Northwestern University's Feinberg School of Medicine, those who sleep more and with a higher quality have more normal heart rates and breathing than those who are sleep deprived (Doss, 2019). I did all my runs after a 2 hours nap after school and with a mean of 6 hours of sleep per day, so the data collected for heart rate is to the best of my knowledge, unaffected by sleep patterns.

*Cardiorespiratory fitness (CRF):* this is the ability of respiratory and circulatory systems to supply oxygen to the skeletal muscles during long physical activity. CRF is measured by oxygen uptake $Vo_{2max}$, when exercising at best effort (Seidenberg & Beutler, 2008). **$Vo_{2max}$** is the maximum amount of oxygen our body can utilize when exercising. The more oxygen you are able to take the faster you become thus the higher the $Vo_{2max}$ the better the running performance. A typical woman has a $Vo_{2max}$ of 27-30 ml/kg/min, while that of and elite female athlete can go as high as 77 ml/kg/min (Capritto,2021). Mine- measured with the same device- is 48 ml/kg/min, hence this factor is controlled as well.

*Stress and anxiety:* it is a known that stress raises the heart rate. My stress levels are usually the highest the days before a test or the day of an important deadline. That is why I did not go for the data collection the days before exams and the day of the deadlines. Doing so, I tried to control stress as much as possible.

*Body temperature and humidity:* When too warm or too cold body tries to compensate for the undesired weather by increasing blood flow when you feel cold and losing heat to the environment when you feel hot. This happens below and above certain temperatures. Where I live there is not much fluctuation in the temperature, the temperature was between 7 and 16 degrees Celsius in the period of data collection, meaning neither too high nor too low, hence controlled.

*The surface one runs on:* when running uphill heart rate increases and when running downhill, it decreases. Moreover, running trails, mountains, pavements and tracks activates different muscles in the body and thus alters heart rates. I chose a flat pavement and kept running at the same place, thus controlling the effect of terrain on Heart rate.

*Medication:* from my knowledge of biology, I know that some of the cough, Thyroid medicine, anti-depressants and supplements can alter and either increase, control or decrease heart rate. As I am taking migraine medications, my heart rate should be higher than normal but as I am taking it every day it is controlled and kept constant throughout the period of data collection.

*Dehydration:* According to NHS one should drink about 1.2 liters- between 6 to 8 glasses- of water to prevent dehydration, however in the hot climate more water is needed ("The Eatwell Guide", 2021). When dehydrated the amount of blood circulating in the body decreases and hence the heart rate increases. I kept the amount of water constant during the days of the data collection and drank 9 glasses per day, 7 before the run and 2 after it, thus dehydration is controlled as well.

*Insufficient recovery after intense work outs:* if we do not take time off sports long enough after a hard workout to recover our heart rate increases for the next workout. That is why I collected the data every other day and did not include any long runs, anything more than 5 miles, in my running plan during the period of data collection. To make sure I have recovered I also used the option on my watch which told me -based on the HR, workout intensity and other factors- when I needed to recover after runs.

*Illness or symptoms of illness:* As the data was collected during pandemic and the season autumn there was a high probability that I would get infected by the COVID-19 or a common flu, that is why I carefully screened for illnesses and made sure that I ran and collected the data only when I felt completely healthy.

*Insufficient nutrition:* when one does not eat enough calories and eats less than the amount of food their body needs their heart rate increases. According to NHS women's recommended calorie intake is 2000 while that of men's is 2500. I tracked my calories and between 2000 to 2300 calories a day during the period of study to control and prevent insufficient nutrition ("What should my daily intake of calories be?", 2021).

*Weight:* Studies show that those who weigh higher, run slower and those who weigh less, run faster. For instance, losing 5 to 10% of body weight improves runners' pace by 3.1 to 5.2% (Zacharogiannis, 2017). This shows that both weight and weight loss affect the pace of a runner, that is why I weighed myself every other day – the days of the data collection. I decided to stop the data collection if I had either gained or lost more than one kilogram.

Less than one kilogram of weight loss and weight gain can be associated with factors such as taking food high in salt, menstruation cycles, weight training etc. during the course of the data collection, the maximum mass I had lost was 600 grams and the maximum I had gained was 500 grams that is why the data collected is unaffected by weight and weight-loss.

***Nutrition, to be more accurate, Carbohydrate intake:*** runners are always advised to eat a lot of carbohydrates at the night before a race to become faster. This shows that there is a relationship between carbohydrate intake and running pace. In order to control the amount of carbohydrates, I did not take any meals high in carbohydrate before or during the day of a run throughout the course of the data collection, that includes meals such as pasta, waffles, rice, potatoes and oats.

***Training level:*** As a runner trains for longer and more efficiently their pace becomes better, they can run one kilometer in shorter amount of time. At the same time with more training heart grows in size, pumps blood more effectively. It can be seen that training level or fitness level affect both heart rate and pace. As the data is just collected based on my runs and in a short period of time during which the body won't change drastically, the training level is kept constant.

**Limitations of data collection:** I tried to control for most of the confounding variables hence it can be stated that the data collection is well, but it cannot be denied that outdoor measurements are less accurate than indoor measurements on the treadmill because GPS measurements are less accurate in the outdoors. Moreover, confounding variables such as wind, altitude, underlying medical conditions were not considered and controlled in this IA.

### Raw results:

In the table below, the pace and heart beats of minutes one to five are depicted, the complete table of raw data which includes the data for 10 minutes and 12 replications can be found in the appendices (see appendix 1). A total of 19 replications were done; however, only 12 of them were taken into account because in the other 7, the data collection had limitations such as malnutrition and symptoms of illness, so it could not be used.

Table 1 Contains the raw data for the first five minutes of 12 days of running. Pace has the unit minute: second and heart rate has the unit beats per minute.

| Pace (1') | Heart rate (1') | Pace (2') | Heart rate (2') | Pace (3') | Heart rate (3') | Pace (4') | Heart rate (4') | Pace (5') | Heart rate (5') |
|---|---|---|---|---|---|---|---|---|---|
| 10.50 | 131.92 | 9.75 | 130 | 9.60 | 135 | 10.76 | 135 | 10.90 | 136 |
| 11.21 | 139 | 10.51 | 151 | 14.66 | 144 | 13.83 | 142 | 13.96 | 141 |
| 14.14 | 124 | 12.83 | 124 | 10.95 | 134 | 10.81 | 132 | 14.06 | 129 |
| 10.90 | 131 | 10.68 | 134 | 10.81 | 137 | 10.71 | 132 | 10.70 | 151 |
| 10.30 | 137 | 11.40 | 143 | 9.36 | 144 | 8.85 | 145 | 12.85 | 139 |
| 8.66 | 139 | 10.56 | 136 | 9.56 | 140 | 11.96 | 144 | 11.51 | 139 |
| 8.55 | 140 | 9.75 | 156 | 10.51 | 131 | 10.56 | 137 | 12.66 | 144 |
| 10.63 | 155 | 10.56 | 156 | 11.23 | 157 | 10.70 | 155 | 9.95 | 153 |
| 12.40 | 138 | 12.40 | 132 | 10.51 | 137 | 11.83 | 136 | 10.10 | 146 |
| 8.85 | 129 | 9.00 | 132 | 9.98 | 127 | 9.70 | 133 | 10.83 | 133 |
| 13.70 | 147 | 10.26 | 147 | 8.56 | 134 | 7.83 | 123 | 7.83 | 138 |
| 10.20 | 141 | 9.50 | 142 | 10.51 | 139 | 11.70 | 173 | 12.65 | 168 |

Pace is written as Minute: Second in the Garmin application; however, as it would not be possible to insert pace in such format and to plot a graph in this way, the seconds were converted into decimals. This was done by multiplying the second by a hundred and dividing the number found by 60. A sample conversion is shown below. Pace shown by Garmin forerunner 945: 10:45, $45 \times 100 = 4500$, $4500 \div 60 = 75$, Thus, the number can be written as 10.75. The numbers in this format were used in this IA.

| Time (minutes) | Average pace | Average heart beat |
|---|---|---|
| 1 | 11.17 | 137.66 |
| 2 | 10.6 | 140.25 |
| 3 | 10.52 | 138.25 |
| 4 | 10.77 | 140.58 |
| 5 | 11.5 | 143.08 |
| 6 | 11.22 | 136.41 |
| 7 | 10.8 | 142.33 |
| 8 | 10.77 | 141.5 |
| 9 | 11.05 | 142.16 |
| 10 | 11.37 | 145.33 |

Table 2 Shows the average heart rates and paces for the first ten minutes and for twelve replications.

## Finding the functions

To find the relation between pace and heart rate, I primarily found the relationships of pace per time and heart rate per time. I kept the time constant and measured heart rate and pace every minute for a total of ten minutes, I then found the relationship of heart rate and pace. There were a total of twelve repetitions in order to increase the reliability of the model found.

Steps:

1- Finding the function of heart beat per time: by experience I knew that when running, as time passes the body becomes tired and the load of exercise gets higher; that is why I thought that time should affect the heart rate, and as more time passes from the time one has started a run, the heart rate should get higher. To find out if that is true, initially the average of the heart rates for each minute were found. There were 12 repetitions thus the mean of the 12 heart rates for each minute were found. Means were used because when the data collection is done using GPS and wearable devices the results of one or two days may not be sufficient and there may be errors in the results, but when this is repeated a number of times the errors will be less and the data collection will be more reliable. After the previous step, the numbers were inserted into the GeoGebra and the scatter plot of heart rate over time was drawn. The graph showed a general increasing trend, meaning as time increased heart beat increased. By looking at the graph, I hypothesize that the function of heart beat per time should have a polynomial model.

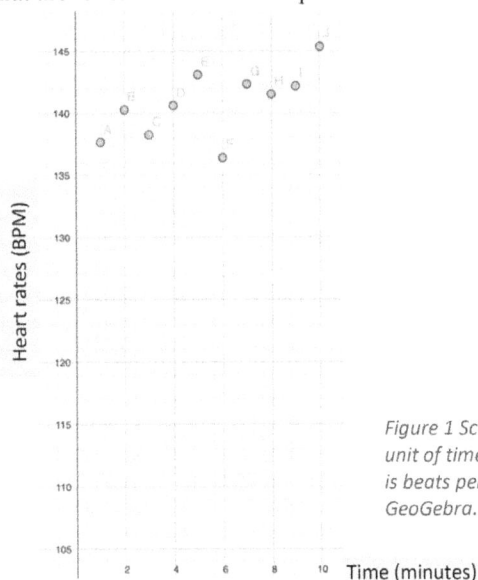

*Figure 1 Scatter plot of heart beat per time; the unit of time is minutes and the unit of heart beat is beats per minute. The Graph is drawn by GeoGebra.*

2- Finding the function of pace per time:  means of the paces of each minute were calculated; as there were 12 repetitions, the mean consisted of 12 numbers. The scatter plot depicting the relationship between pace and time was drawn and a general increasing trend was observed; this function can be a sine function or it could show a polynomial model, I will discuss both these models in the following pages.

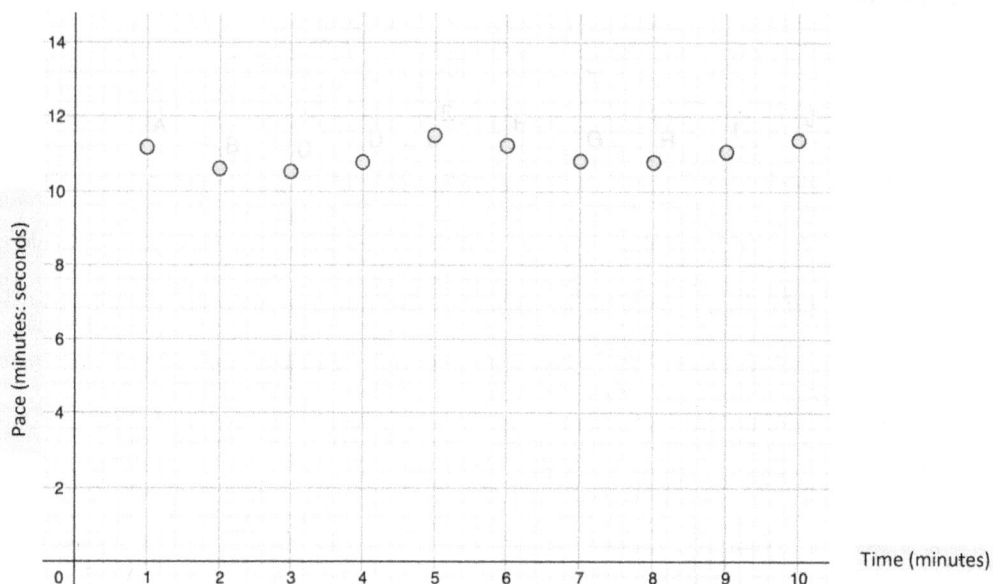

*Figure 2 Scatter plot of pace per time; the unit of time is minutes and the unit of pace is minutes: seconds. The graph is drawn by GeoGebra.*

## Mathematical modelling and interpretation of results:

To find the relationship between pace and heart rate I initially found out which models are the best for showing the relationship between pace per time and heart rate per time. Subsequently, I chose the best models and I used the chain rule discussed in the higher-level mathematics to merge the two relationships and find the function of heart rate per pace.

## Pace per time:

***Trigonometric model (sine function):*** sine function has the general formula: $y = asin(b(x - c)) + d$ in this function a is the amplitude which should be written as $|a|$, d indicates the vertical translation, c indicates the horizontal translation, the period is determined by b, it should be found in this way: $\frac{2\pi}{b}$ for b > 0.

The aforementioned function is obtained from a horizontal stretch by $\frac{1}{b}$ and a vertical stretch by a, followed by a translation of $\binom{c}{d}$ $(Cirrito \,\& \, Tobin, 2004)$ .

To find this function by hand the following steps were performed:

- To find the altitude of the sine function we need to subtract the smallest y from the greatest y and then divide the number found by two

$$a = \frac{11.50 - 10.52}{2} = 0.49$$

- To find the vertical shift of the sine function we have to subtract the altitude from the highest y value.

$11.50 - 0.49 = 11.01$

- The period of a sine function is found by the formula: $\frac{2\pi}{b}$, for this particular function the b is 5, thus the period is $\frac{2\pi}{5}$

- To find the horizontal shift of this function, we have to put the coordinates of one of the points in the equation and set c as an unknown. Any of the points can be chosen for this purpose. In this IA, all of the points were put inside the equations, 10 c values were found and the corresponding graphs were sketched. The value of the c which gave the closest graph to the GeoGebra graph was chosen to be used in the final function. Below an example of this process will be shown.

$$m = 11.01 + 0.49 \sin\left(\frac{2\pi}{5}(X - C)\right)$$

$(10, 11.37)$ this point will be plugged into the function as an example.

$$11.01 + 0.49 \sin\left(\frac{2\pi}{5}(10 - C)\right) = 11.37$$

$$\sin\left(\frac{2\pi}{5}(10 - C)\right) = \frac{11.37 - 11.01}{0.49}$$

$$\sin\left(\frac{2\pi}{5}(10 - C)\right) = 0.734$$

$$\frac{2\pi}{5}(10 - C) = \sin^{-1}(0.734)$$

$$C = 13.15$$

Function found by mathematical methods: $m(x) = 11.01 + 0.49 \sin\left(\frac{2\pi}{5}(x - 13.15)\right)$

Function given by GeoGebra:

$$d(x) = 11 + 0.47 \sin(1.22X + 1.55)$$

graph drawn by GeoGebra:

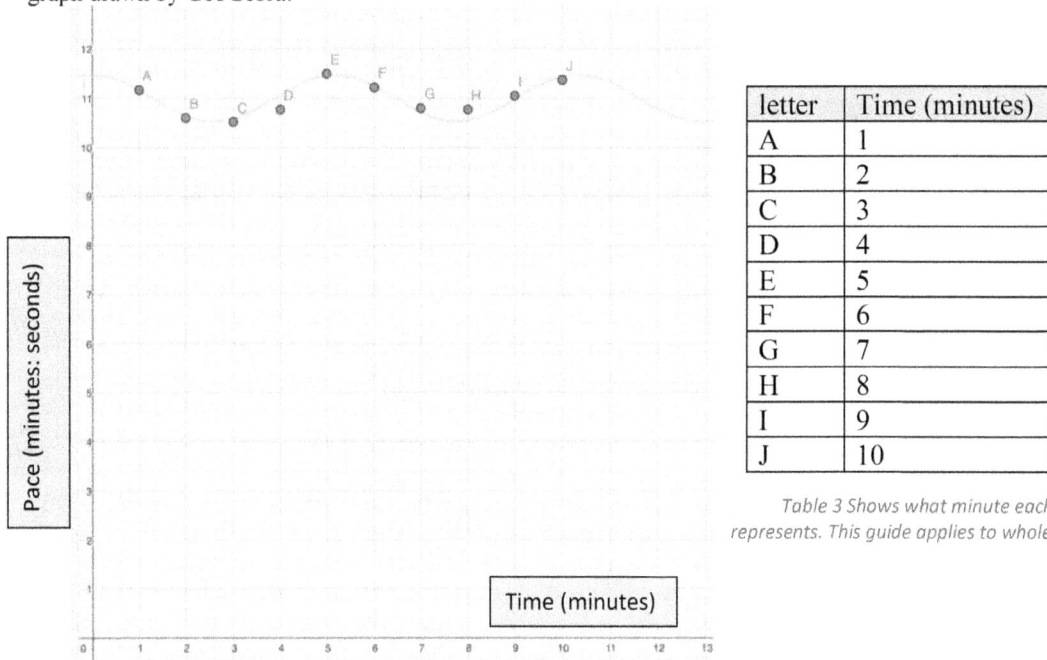

| letter | Time (minutes) |
|--------|----------------|
| A | 1 |
| B | 2 |
| C | 3 |
| D | 4 |
| E | 5 |
| F | 6 |
| G | 7 |
| H | 8 |
| I | 9 |
| J | 10 |

*Table 3 Shows what minute each letter represents. This guide applies to whole IA.*

*Figure 3 Model of Pace Vs. time. The unit for time is minutes and the unit for pace is minutes: seconds.*

The c which gave a graph closest to the GeoGebra graph was C=13.15, below the graph drawn with the equation found by hand and the graph drawn by the graphics calculator are compared. The comparison shows that the trigonometric model found by me is in very good accordance with the model found by the application GeoGebra.

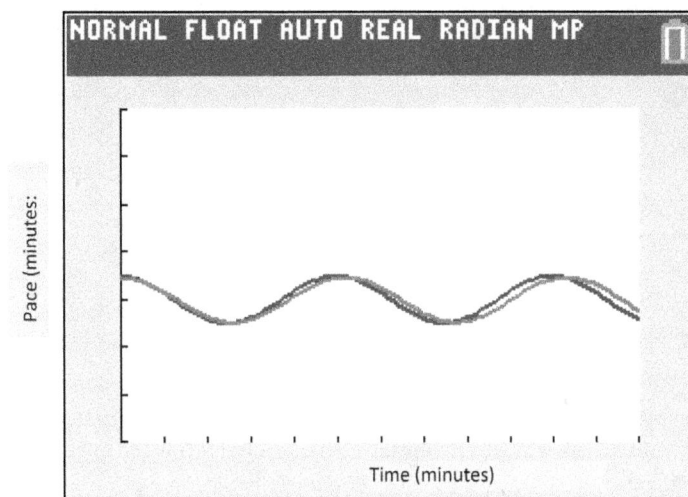

```
NORMAL FLOAT AUTO REAL RADIAN MP
```

*Figure 4 The red graph is the graph drawn by the calculator and the blue graph is the graph drawn by hand. the Blue graph has equation $m= 11.01 + 0.49\ sin\left(\frac{2\pi}{5}(x - 13.15)\right)$ and the red graph has the equation $d=11 + 0.47\ sin(1.22X + 1.55)$. The application ti-connect was used to transfer the graphs from the calculator to the computer.*

The next model I want to discuss is the polynomial model; I will make use of matrices- discussed in the HL syllabus- to find this model.

### *Polynomial model:*

A matrix is a rectangular array of numbers arranged in p columns and b rows. It is shown with a capital letter and it has orders. The order of a matrix with b rows and p columns should be written as $b \times p$. When b and p are equal it can be said that the matrix M is a square matrix of order b. When matrix M has order $b \times p$, it should be written as:

$M = \left(a_{i_j}\right)$ where $i = 1,2,3, \ldots, m$ & $j = 1,2,3, \ldots, n$ & $a_{i_j}$ is the element in the $i$th row and $j$th column

(Economopoulos et al., 2019).

Matrix multiplication can help us express and solve systems of equations. Any system with n variables and m equations can be written in the form of a matrix equation. For example:

$$f(x) = ax^2 + bx + x$$

147

$$y_1 = ax_1{}^2 + bx_1 + C$$

$$y_2 = ax_2{}^2 + bx_2 + C$$

$$y_3 = ax_3{}^2 + bx_2 + C$$

$$\begin{bmatrix} x_1^2 & x_1 & 1 \\ x_2^2 & x_2 & 1 \\ x_3^2 & x_3 & 1 \end{bmatrix} \times \begin{bmatrix} a \\ b \\ c \end{bmatrix} = \begin{bmatrix} y_1 \\ y_2 \\ y_3 \end{bmatrix}$$

A m × m system of equations can be written in the following format:

$$AX = B$$

In this way, A is a m × m matrix which contains the coefficients and B is the m × 1 matrix which contains the coefficients and X is a m × 1 matrix containing the variables. We then want to determine X in the following way:

$$Multiply\ A^{-1}from\ left\ side.$$

$$AB \neq BA$$

$$A^{-1}(AX) = A^{-1}B \Rightarrow A^{-1}AX = A^{-1}B$$

$$A^{-1}A = I\ and\ IX = X$$

$$IX = A^{-1}B \Rightarrow X = A^{-1}B$$

When I was finding the function which suited the data the best, I noticed that a number of different polynomial models with different degrees, including different points, fitted the data pretty well; however, I had to choose the best of these and that is why other factors were also taken into consideration. Different functions and their respective graphs which were relatively good predictors of pace Vs. time are available in the appendices (see appendix2); these functions were analyzed and the best one was chosen. One important factor was that when the running had not started (time zero), g(x) and f(x) had pace values which were relatively high (12.2 and 14.1 respectively) and this was not possible because at that point, the running had not even started yet and the pace should have been as low as possible. On the other hand, both p(x) and g(x) seemed to be good predictors of the pace for the first ten minutes; however, p(x) included more points on its graph so it was chosen as the final function.

$$p(x) = ax^7 + bx^6 + cx^5 + dx^4 + ex^3 + fx^2 + gx + h$$

A: $11.17 = a(1)^7 + b(1)^6 + c(1)^5 + d(1)^4 + e(1)^3 + f(1)^2 + g(1) + h$

B: $10.6 = a(2)^7 + b(2)^6 + c(2)^5 + d(2)^4 + e(2)^3 + f(2)^2 + g(2) + h$

C: $10.52 = a(3)^7 + b(3)^6 + c(3)^5 + d(3)^4 + e(3)^3 + f(3)^2 + g(3) + h$

F: $11.22 = a(6)^7 + b(6)^6 + c(6)^5 + d(6)^4 + e(6)^3 + f(6)^2 + g(6) + h$

G: $10.8 = a(7)^7 + b(7)^6 + c(7)^5 + d(7)^4 + e(7)^3 + f(7)^2 + g(7) + h$

H: $10.77 = a(8)^7 + b(8)^6 + c(8)^5 + d(8)^4 + e(8)^3 + f(8)^2 + g(8) + h$

I: $11.05 = a(9)^7 + b(9)^6 + c(9)^5 + d(9)^4 + e(9)^3 + f(9)^2 + g(9) + h$

J: $11.37 = a(10)^7 + b(10)^6 + c(10)^5 + d(10)^4 + e(10)^3 + f(10)^2 + g(10) + h$

$$p(x)= \begin{Vmatrix} 1 & 1 & 1 & 1 & 1 & 1 & 1 & 1 \\ 128 & 64 & 32 & 16 & 8 & 4 & 2 & 1 \\ 2187 & 729 & 243 & 81 & 27 & 9 & 3 & 1 \\ 78125 & 15625 & 3125 & 625 & 125 & 25 & 5 & 1 \\ 823543 & 117649 & 16807 & 2401 & 343 & 49 & 7 & 1 \\ 2.1E6 & 262144 & 32768 & 4096 & 512 & 64 & 8 & 1 \\ 4.78E6 & 531441 & 59049 & 6561 & 729 & 81 & 9 & 1 \\ 1E7 & 1E6 & 100000 & 10000 & 1000 & 100 & 10 & 1 \end{Vmatrix} \times \begin{bmatrix} a \\ b \\ c \\ d \\ e \\ f \\ g \\ h \end{bmatrix} =$$

$$\begin{bmatrix} 11.17 \\ 10.6 \\ 10.52 \\ 11.22 \\ 10.8 \\ 10.77 \\ 11.05 \\ 11.37 \end{bmatrix} = AX = B \quad \Rightarrow \quad X = B \times A^{-1}$$

$$X = \begin{bmatrix} 1.92 \times 10^{-4} \\ -0.007 \\ 0.13 \\ -1.11 \\ 5.11 \\ -12.13 \\ 13.18 \\ 5.99 \end{bmatrix}$$

Function found by the matrix method:

$$p(x) = (1.92 \times 10^{-4})x^7 + (-0.007)x^6 + 0.13x^5 + (-1.11)x^4 + (5.11)x^3 + (-12.13)x^2 + (13.18)x + 5.99)$$

Since the matrix has to be a square matrix for the calculator to work, points D and E- which were the furthest from the line on the graph- had to be disregarded in the matrix method calculated by me. In the GeoGebra model, point D was disregarded as it acted like an outlier in comparison to the other points.

Function found by GeoGebra: $a(x) = 0x^7 - 0.01x^6 + 0.13x^5 - 1.12x^4 + 5.11x^3 - 12.11x^2 + 13.15x + 6.01$

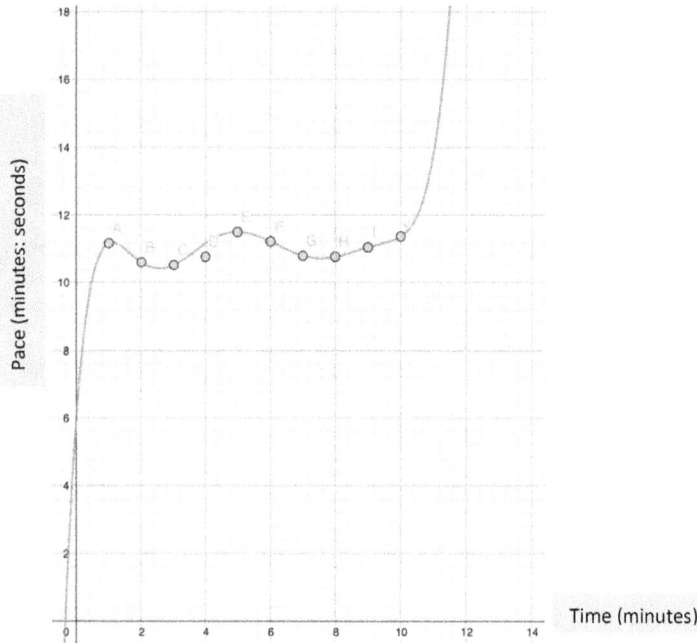

*Figure 5 Pace Vs. time, time has the unit minutes and pace has the unit (minute: seconds).*

The polynomial model found by the matrix method had some differences with the model found by GeoGebra; these differences might have happened due to several reasons. Firstly, for the matrix method, the dimensions of the matrix had to be equal to each other, so for a degree 7 model, a $8 \times 8$ matrix (or in mathematical language a square matrix of order 8) should have been used, hence 2 of the points which were the furthest from the best fit line had to be eliminated but in the GeoGebra model only one point which was the furthest from the best fit line had to be eliminated. The second issue was that calculators have less accuracy than the GeoGebra application, this is even seen when plotting graphs during the IB classes. Overall, these differences did not make the function an unreliable one and the polynomial model still included more points in it than the sine model, thus it was used for the rest of the analysis as well.

Heart rate per time:

The model for heart rate per time was polynomial model with a degree of 7. Below the mathematical steps are

explained.

*general equation:* $b(x) = ax^7 + bx^6 + cx^5 + dx^4 + ex^3 + fx^2 + gx + h$

$A: 137.66 = a(1)^7 - b(1)^6 + c(1)^5 - d(1)^4 + e(1)^3 - f(1)^2 + g(1) + h$

$B: 140.25 = a(2)^7 - b(2)^6 + c(2)^5 - d(2)^4 + e(2)^3 - f(2)^2 + g(2) + h$

$C: 138.25 = a(3)^7 - b(3)^6 + c(3)^5 - d(3)^4 + e(3)^3 - f(3)^2 + g(3) + h$

$E: 143.08 = a(5)^7 - b(5)^6 + c(5)^5 - d(5)^4 + e(5)^3 - f(5)^2 + g(5) + h$

$G: 142.33 = a(7)^7 - b(7)^6 + c(7)^5 - d(7)^4 + e(7)^3 - f(7)^2 + g(7) + h$

$H: 141.5 = a(8)^7 - b(8)^6 + c(8)^5 - d(8)^4 + e(8)^3 - f(8)^2 + g(8) + h$

$I: 142.16 = a(9)^7 - b(9)^6 + c(9)^5 - d(9)^4 + e(9)^3 - f(9)^2 + g(9) + h$

$J: 145.33 = a(10)^7 - b(10)^6 + c(10)^5 - d(10)^4 + e(10)^3 - f(10)^2 + g(10) + h$

$b(x)=$

$$\begin{Vmatrix} 1 & 1 & 1 & 1 & 1 & 1 & 1 & 1 \\ 128 & 64 & 32 & 16 & 8 & 4 & 2 & 1 \\ 2187 & 729 & 243 & 81 & 27 & 9 & 3 & 1 \\ 279936 & 46656 & 7776 & 1296 & 216 & 36 & 6 & 1 \\ 823543 & 117649 & 16807 & 2401 & 343 & 49 & 7 & 1 \\ 2.1E6 & 262144 & 32768 & 4096 & 512 & 64 & 8 & 1 \\ 4.78E6 & 531441 & 59049 & 6561 & 729 & 81 & 9 & 1 \\ 1E7 & 1E6 & 100000 & 10000 & 1000 & 100 & 10 & 1 \end{Vmatrix} \times \begin{bmatrix} a \\ b \\ c \\ d \\ e \\ f \\ g \\ h \end{bmatrix} =$$

$$= \begin{vmatrix} 137.66 \\ 140.25 \\ 138.25 \\ 143.08 \\ 142.33 \\ 141.5 \\ 142.16 \\ 145.33 \end{vmatrix} = Ax = B \Rightarrow X = B \times A^{-1}$$

$$\Rightarrow X = \begin{bmatrix} 8.579 \times 10^{-4} \\ -0.037 \\ 0.652 \\ -6.012 \\ 30.63 \\ -84.217 \\ 112.98 \\ 83.655 \end{bmatrix}$$

Function found by the matrix method:

$b(x) = (8.579 \times 10^{-4})x^7 + (-0.037)x^6 + (0.652)x^5 + (-6.012)x^4 + (30.63)x^3 + (-84.21)x^2 + (112.98x) + (83.655)$

Function found by GeoGebra:

$r(x) = 0x^7 - 0.04x^6 + 0.65x^5 - 6.01x^4 + 30.63x^3 - 84.22x^2 + 112.98x + 83.66$

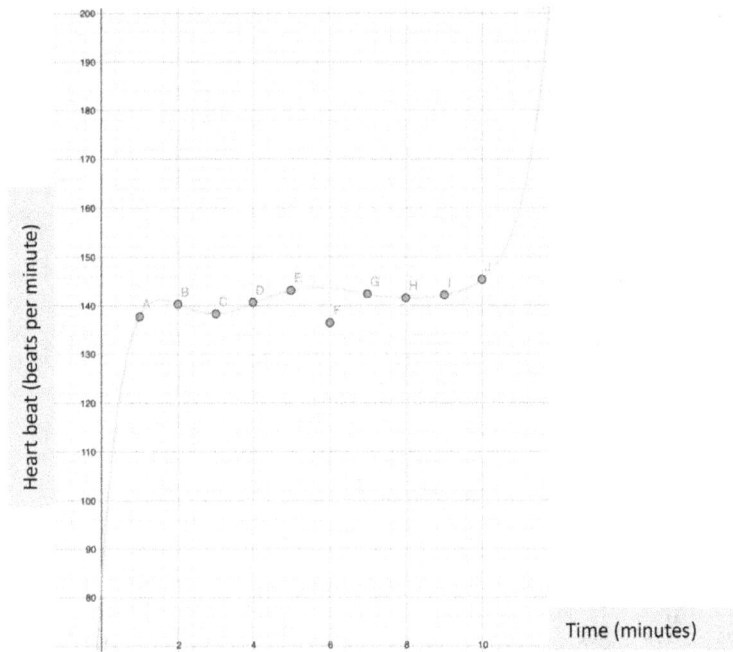

*Figure 6 Heart rate Vs. time. Heart rate has unit beats per minutes and time has unit minutes.*

By looking at the graph of the polynomial model for Heart beat per time it can be seen that after ten minutes the regression line is increasing drastically. This shows that the model cannot be used for any time after 10 minutes. The last step was to converge the two functions derived and to find the relationship between pace and heart rate.

### *The equation of heart rate per pace:*

Chain rule or the "Outside-Inside" rule:
The chain rule discussed in the HL syllabus can be written in the following way:

$$(f \circ g)'(x) = f'(g(x)) \cdot g'(x)$$

$$\frac{dy}{dx} = \frac{dy}{du} \times \frac{du}{dx} \qquad \text{Formula 1}$$

$$\frac{dy}{dx} = f'(g(x)) \times g'(x)$$

$$\frac{d}{dx}f(u) = f'(u)\frac{du}{dx}$$

In this method you have to initially differentiate the outside function $f$, the then evaluate it at the inside function $g$ which has been left alone and multiply the whole thing by the derivative of the inside function (Hass et al., 2018, p. 143)

The final equation using the two functions derived from the matrix method:

$$p(x) = (1.92 \times 10^{-4})x^7 + (-0.007)x^6 + 0.13x^5 + (-1.11)x^4 + (5.11)x^3 + (-12.13)x^2 + (13.18)x + 5.99)$$

$$b(x) = (8.579 \times 10^{-4})x^7 + (-0.037)x^6 + (0.652)x^5 + (-6.012)x^4 + (30.63)x^3 + (-84.21)x^2 + 112.98x + 83.655$$

$$\frac{dp}{dx} = p'(x) = (13.44 \times 10^{-4})x^6 - 0.042x^5 + 0.65x^4 - 4.44x^3 + 15.33x^2 - 24.26x + 13.18$$

$$\frac{db}{dx} = b'(x) = 60.05 \times 10^{-4}x^6 - 0.22x^5 + 3.26x^4 - 24.04x^3 + 91.89x^2 - 168.42x + 112.98$$

Using formula 1 which was explained above, the two functions get merged.

$$\frac{db}{dp} = \frac{db}{dx} \times \frac{dx}{dp}$$

$$\frac{db}{dp} = (60.05 \times 10^{-4}x^6 - 0.22x^5 + 3.26x^4 - 24.04x^3 + 91.89x^2 - 168.42x + 112.98)$$

$$\times \left(\frac{1}{(13.44 \times 10^{-4})x^6 - 0.042x^5 + 0.65x^4 - 4.44x^3 + 15.33x^2 - 24.26x + 13.18}\right)$$

Note that the function p(x) depicts the relationship between pace and time, while the function b(x) represents the relationship between heart rate and time.

The final equation using the two functions derived from GeoGebra:

$$a(x) = 0x^7 - 0.01x^6 + 0.13x^5 - 1.12x^4 + 5.11x^3 - 12.11x^2 + 13.15x + 6.01$$

$$r(x) = 0x^7 - 0.04x^6 + 0.65x^5 - 6.01x^4 + 30.63x^3 - 84.22x^2 + 112.98x + 83.66$$

$$\frac{da}{dx} = a'(x) = x^6 - 0.06x^5 + 0.65x^4 - 4.48^3 + 15.33x^2 - 24.22x + 13.15$$

$$\frac{dr}{dx} = r'(x) = x^6 - 0.24x^5 + 3.25x^4 - 24.04x^3 + 91.89x^2 - 168.44x + 112.98$$

$$\frac{dr}{da} = (x^6 - 0.24x^5 + 3.25x^4 - 24.04x^3 + 91.89x^2 - 168.44x + 112.98) \times$$

$$\left(\frac{1}{x^6 - 0.06x^5 + 0.65x^4 - 4.48^3 + 15.33x^2 - 24.22x + 13.15}\right)$$

Note that the function a(x) depicts the relationship between pace and time, while the function r(x) represents the relationship between heart rate and time.

## *Conclusion*

The aim of this paper was to find the relationship between pace and heart rate. This was successfully done by controlling a number of confounding variables, recording the data for 12 days, finding the means of heart rate and pace in each of the ten minutes, deriving separate functions for heart rate per time and pace per time and finally converging them into one using the chain rule. The hypothesis was that heart rate should increase as the pace increased; however, the relationship found did not show a general increasing or decreasing trend and was much more complex than the hypothesis. The final function derived was a rational function.

A limitation of both the pace per time and heart rate per time functions found by the matrix methods was that as the calculator had less accuracy and as some minutes had to be disregarded in finding the two functions, they were less accurate than the functions found by GeoGebra and they were not able to predict some minutes correctly. Since the final function was also derived from the two separate functions, the same problem is also present in the final function. If we compare the functions $\frac{db}{dp}$ $and$ $\frac{dr}{da}$, it can be seen that the difference is not really high and it is negligible. The accuracy issue; however, seems to be natural when the modelling is not done via computer applications with higher accuracy and memory such as GeoGebra.

By looking at the models of heart beat per time and pace per time, it could be seen that the models were not applicable to time periods after ten minutes. The concept of heart rate, its fluctuations and how it varies during running is of the most complex and debatable topics of the days, that is why, it should not be unexpected that a mathematical function would not be able to predict the heart rate based on pace for all periods of time.

Overall, the final function found is acceptable and does show the relationship between pace and heart rate in the first ten minutes of running, with more considerations and use of more accurate devices a better model can be found.

An extension to this IA would be modelling heart rate per pace for other periods of time, for example for ten to twenty minutes or other periods of time. Another point is that as many different variables impacted the independent and dependent variables, pace and heart rate respectively, this model is only applicable to the individual who collected this first-hand data. If one is interested in further exploring this topic, they can recruit a high number of athletes, control the variables, gather data from their training sessions, group them into beginner, intermediate and advanced runners and then create a more complex model which can work for all people. Even though this requires a lot of time and proficiency, it can still be done by high school students in most schools which have sports teams.

To improve the current IA, one can think of controlling more of the confounding variables which may have affected the data collection; factors such as wind, altitude, time of the day one runs at, medical conditions and the precision and accuracy of the instrument used for data collection are some examples which could be regarded as well. One can also repeat the data collection more times and increase the reliability of the functions by this method.

# 6. THE FOURIER TRANSFORM AND ITS USES IN SOUND ANALYSIS

Author: Joseph Azrak
Moderated Mark: 20/20
Level: Math AA HL

# 1 Introduction

Mathematical modeling has been of utmost interest to me for some time. I have always been intrigued by the real-life applications of mathematics in domains ranging from physical phenomena to modeling of human societies. However, the mathematics of sound has been of great interest to me recently. It is fairly simple to describe sound at an elementary level—sound is, at its core, a *periodic variation* in air pressure. An emitting device, such as a speaker or one's vocal cords, periodically displaces air molecules back and forth. These molecules displace the molecules next to them, and so continues the chain, creating what is called a *longitudinal wave*. Sound is especially important to humans. Music can inspire awe, happiness, and melancholy. Hearing a loved one's voice or a beautiful quartet can produce a neurochemical cascade of intense emotion. It is, then, truly remarkable that the perceived diversity of sound in our everyday experience is merely the product of variation in periodicity in sound pressure. To me, this seemed somewhat incredulous. After all, consider the sheer number of ways we describe sound in everyday life. A sound can be *coarse*, *smooth*, *melodic*, *cringeworthy*, perhaps *mellow*. Some sounds are considered *bright*, others are considered *warm*, some *hearty*. The human voice has acoustic qualities that clearly separate it from a guitar string, but yet it's difficult to pinpoint what exactly is different. Indeed, we even differentiate between humans based on their voice without knowing exactly how we do so. We can describe and classify the phenomenon of sound in so many different ways that it seems infeasible to do so mathematically. This deeply intrigued me and inspired me to investigate the phenomenon of sound and the underlying trigonometry, complex numbers, and calculus. In this investigation, I will explain what makes sounds unique and how one of the most impactful algorithms today—the Fourier transform—operates. I will investigate the square wave using the transform and pull interesting mathematical insight. Further, I will apply the transform to example data and then detail my personal endeavour to record real sound data and analyse it with a programmed Fourier transform.

# 2   The mathematics of sound

As stated previously, sound is, at its core, a periodic variation in air pressure. The sine function, denoted $\sin x$, is perhaps the most widely used mathematical model of periodic behavior. It is a smooth, repeating curve, with a well-defined amplitude and period.

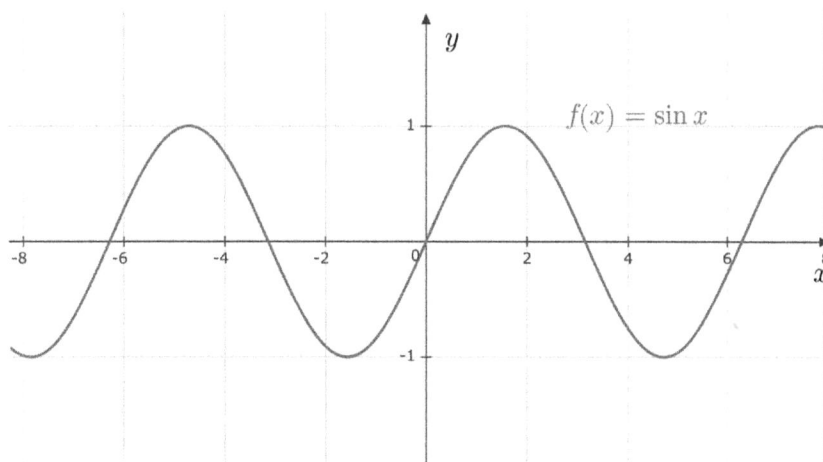

Figure 1: A pure sine wave defined by $f(x) = \sin x$. When perceived by a human ear, it sounds like a single tone.

Indeed, if one were to build a device which displaces the air molecules around it at a rate equivalent to that of $y = \sin x$ — for example, by using an electromagnet — a sound would be perceived. In this context of sound, the "height" of the wave is the instantaneous *displacement* of the air molecules at that time. We now introduce a property of sound: the *frequency*. The frequency of a pure sound is well understood colloquially as how "high- or low-pitched" a sound is. Mathematically speaking, such a characterization only applies to *pure tones*: that is, sounds which are the product of a single sine wave (for example, in Figure 1). We can alter the frequency by adding a factor $B$ into the expression $f(x) = \sin Bx$, where $B \propto$ frequency. Finally, the *phase shift* of a sinusoid refers to the left- or right-shift applied to it, if any.

In life, the sounds we hear rarely are pure tones. Almost all the sounds we hear—for

159

example, voices and instruments—are instead a linear sum of *multiple* pure tones, each of a different frequency. When pure tones are added in this way, the resulting sound is called a *harmonic*. The addition of multiple sine waves of different frequencies results in behavior that is much more complex than that of the constituent waves. It is this phenomenon that I am covering in the investigation. It is simple to define the sum of multiple sine waves mathematically. Given two pure tones $f(t)$ and $g(t)$, their resulting harmonic can be defined as simply $K(t) = f(t) + g(t)$. Geometrically, this means that we sum the individual displacements (heights) $y$-values for each time to obtain the new displacements (heights). For example, if $f(t) = 2\sin t$ and $g(t) = 9\sin t$, their sum would be $K(t) = 2\sin t + 9\sin t$. At the third second, for example, the amplitude of the harmonic would be $K(3) = 2\sin 3 + 9\sin 3$. Figure 2 shows increasingly complexity when two and three pure tones are added in this way—complicated waveforms emerge.

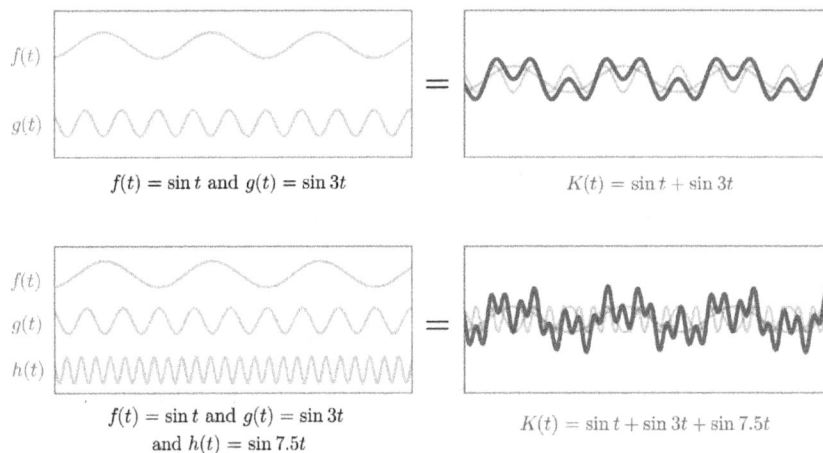

$f(t) = \sin t$ and $g(t) = \sin 3t$

$K(t) = \sin t + \sin 3t$

$f(t) = \sin t$ and $g(t) = \sin 3t$
and $h(t) = \sin 7.5t$

$K(t) = \sin t + \sin 3t + \sin 7.5t$

Figure 2: Two and then three sine waves added result in mounting complexity. The resultant wave is shown in blue.

# 3  Decomposing harmonics

Clearly, as more pure tones are added, complexity arises, and periodicity appears to recede. Make no mistake—periodicity is still there, albeit hidden. A central question in sound analysis has been the process of *reversing* this linear addition. In other words, given a

## 3.2 Derivation of the Fourier transform

The Fourier transform arose from an intuition by Joseph Fourier. While the inner workings of the Fourier transform are not particularly relevant to this this investigation—the practical aspect is—the basic idea will be detailed. Let's imagine a rotating vector $\vec{a}$. We make it so that at every point in time $t$, the magnitude of $\vec{a}$ is $f(t)$—the function we want to decompose. Complex numbers are an elegant mathematical tool to analyse vectors because they are inherently two-dimensional—as such, let's make our vector rotate around a complex plane at some constant angular velocity that we pick $\omega$. This situation is shown in Figure 4. We can see the vector as a pen which is drawing the original function $f(t)$, except it is

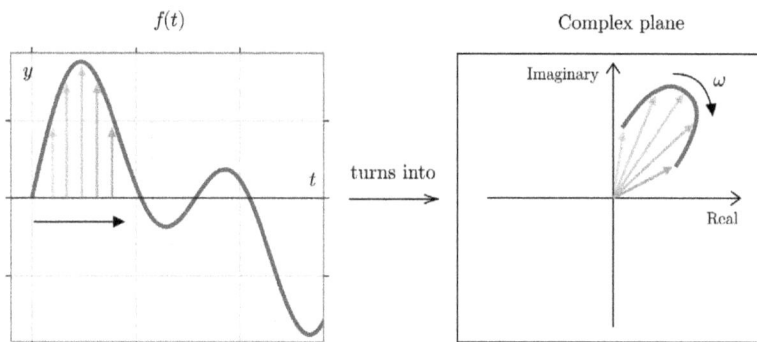

Figure 4: It is as if the vector $\vec{a}$ is a pen "wrapping" $f(x)$ around a circle at varying speeds, or angular velocities, $\omega$.

doing so around a circle and with an angular velocity that we pick $\omega$. We can also talk about the pen's frequency $\zeta = \frac{\omega}{2\pi}$, which is the number of times the pen makes a complete circle per unit time. The shape produced is what I like to call a "Fourier flower". Technically speaking, it is a "complex-plane representation of the original time-domain function", but "Fourier flower" has a better ring to it. Figure 5 shows three examples of Fourier flowers, each of the same origin function, but with varying "pen speeds" $\omega$. We proceed to examine a property of each of these graphs—their so-called "centre of mass". It's difficult to concisely define this property, but some graphs seem less "balanced" than others. Notice that save for Diagram C, all of the diagrams in Figure 5 have a "centre of mass" (shown by the black

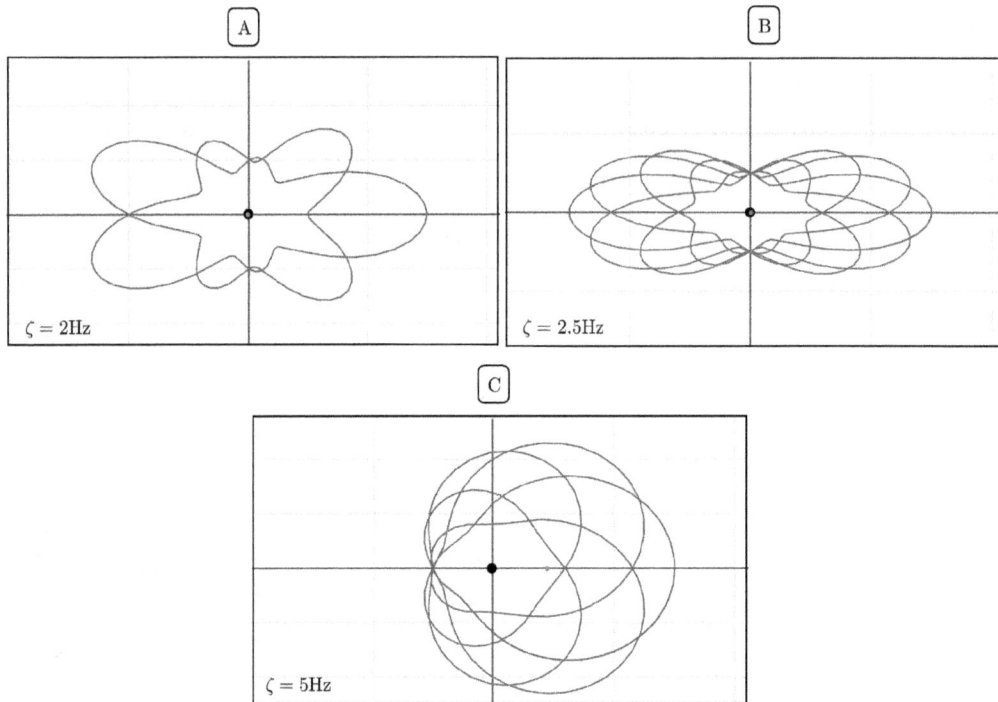

Figure 5: Varying the wrapping frequency produces vastly different shapes. $\zeta = 5$ produces a centre of mass that is unusually far to the right (the red dot is the centre of mass of that flower).

dot) that is relatively balanced at the origin $(0, 0)$. However, for wrapping frequency $\zeta = 5$, the centre of mass is unusually far to the right, and the drawn shape appears unbalanced in that direction. Fourier concluded that this only occurred **when the wrapping frequency coincided with one of the component frequencies of a harmonic**. That is, if we notice that a specific $\zeta$ causes an unbalanced centre of mass, we can isolate that $\zeta$ as being a likely pure tone frequency component of the original function. To do this mathematically, we need to create a model for the "wrapping" of the original function around the complex plane and of the concept of taking the resulting curve's "centre of mass". We can represent the position of the rotating vector at any specific point in time by using trigonometry. Figure 6 shows how we can decompose the vector's components and apply Euler's formula. If the vector's rotational frequency is $\zeta$, its angular position at a time $t$ must be $\theta = 2\pi\zeta t$. We can "stretch" the vector to be the length of the amplitude of the function by setting

162

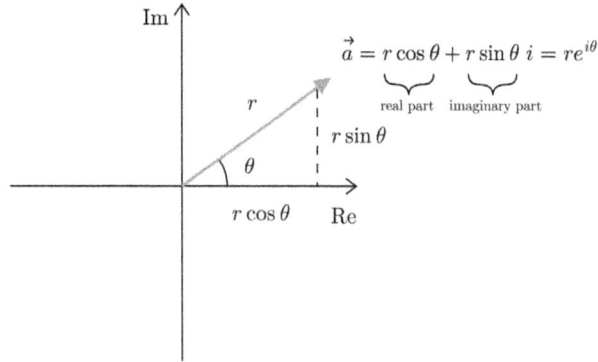

Figure 6: By reasoning geometrically, we arrive to $\vec{a} = r\cos\theta + r\sin\theta i$, which equals $re^{i\theta}$ by Euler's identity.

$r = f(t)$. Thus, at a given point in time, the vector is defined by $\vec{a} = f(t)e^{2\pi i \zeta t}$.

Now that we have a method to mathematically describe the complex position of the rotating vector, we need to take its "centre of mass", or how much it skews to a single direction. This could be done by sampling $N$ points in the set of points and averaging them. However, this is not mathematically pure because our result's resolution would depend on $N$, the sample size. As such, we simply take the integral from negative infinity to infinity of the Fourier flower. A negative sign is included in the exponent as part of convention:

$$\mathscr{F}(\zeta) = \int_{-\infty}^{\infty} f(t)e^{-2\pi i \zeta t}\, \mathrm{d}t \tag{1}$$

This is the definition of the Fourier transform. Of course, $\mathscr{F}(\zeta)$ returns a complex number (the centre of mass of that Fourier flower). That complex number has a modulus and an argument—the modulus (defined by $\sqrt{\mathrm{Re}^2 + \mathrm{Im}^2}$) tells us the abundance of the frequency and the argument tells us its phase-shift.

## 3.3 Investigation of the square wave using Fourier transforms

The square wave, usually denoted $S(t)$, is a standard example of a periodic function. Unlike a sine wave, it has instantaneous transitions from its positive and negative peaks. While the square wave has a formal mathematical definition, it can simply be thought of as the "sign of the sine function" in this paper. An example of a square wave is shown in Figure 7.

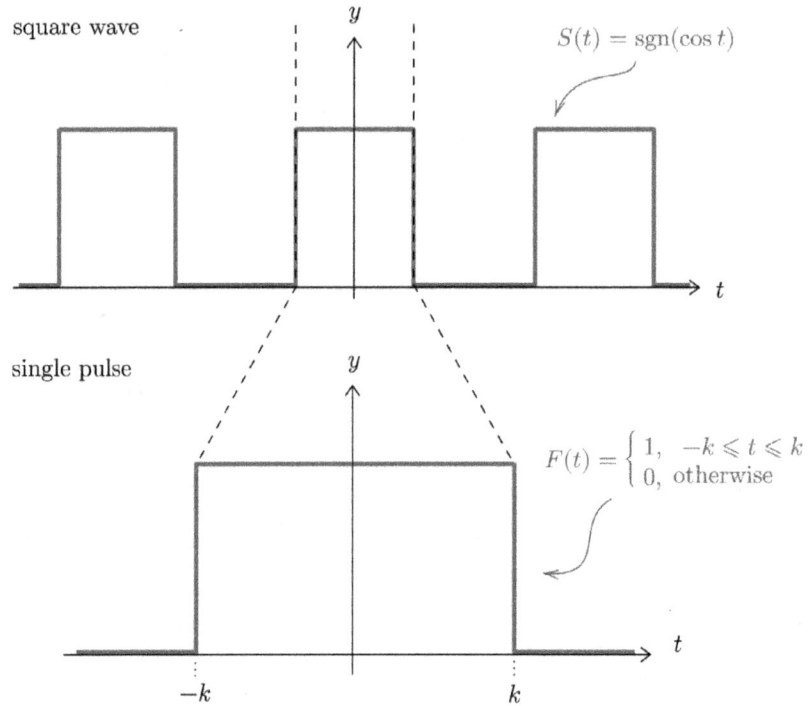

Figure 7: Example of a square wave, as well as the isolation of a single pulse in that wave.

In our analysis, we will not be concerned with the square wave in general. We will narrow our analysis to a single pulse in the wave–this isolation is also shown in Figure 7. Analysis of the whole wave requires Fourier series, which are outside of the scope of this paper. While the square wave does have a mathematical definition, we are only looking at one pulse—as such, we can simply define it as the piecewise function

$$F(t) = \begin{cases} 1, -k \leq t \leq k \\ 0, \text{otherwise} \end{cases}$$

In our analysis, we also assume the wave is shifted up so that it is entirely positive.

We can analyse the square wave as a sum of infinitely many sinusoids–as such, putting it through a Fourier transform should reveal a frequency function $\mathscr{F}(\zeta)$ that reveals component frequencies. Recall that we determined the Fourier transform of $F(t)$ to be:

$$\mathscr{F}(\zeta) = \int_{-\infty}^{\infty} F(t)e^{-2\pi i \zeta t}\mathrm{d}t$$

This troubled me somewhat. I was unclear on how I were to symbolically determine an infinite integral. However, recall the definition of the integral. We define $\int_a^b f(x)\,\mathrm{d}x$ as the *area under the graph* of $f(x)$ bound by a minimum of $x = a$ to a maximum of $x = b$, as shown in Figure 8. However, our function is "flat" for all $t \notin [-k, k]$–that is, the area

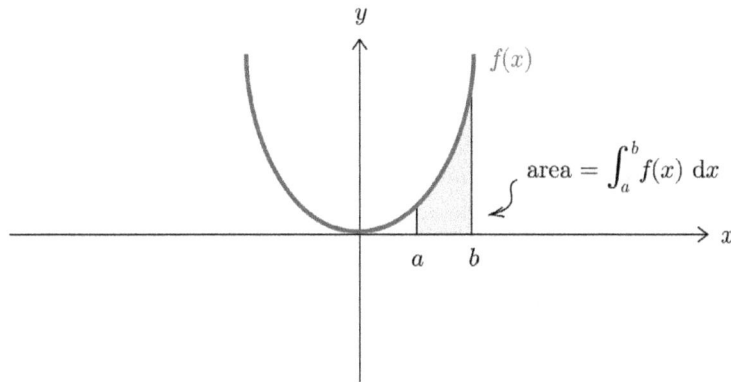

Figure 8: We can find the area under the plot of a function by integrating it.

under is 0. We can then simplify our integral to the bounds $-k$ and $k$, as that is where the "area" is found. In other words,

$$\int_{-\infty}^{\infty} F(t)e^{-2\pi i \zeta t}\,\mathrm{d}t = \underbrace{\int_{-\infty}^{-k} F(t)e^{-2\pi i \zeta t}\,\mathrm{d}t}_{0} + \int_{-k}^{k} F(t)e^{-2\pi i \zeta t}\,\mathrm{d}t + \underbrace{\int_{k}^{\infty} F(t)e^{-2\pi i \zeta t}\,\mathrm{d}t}_{0}$$

Also recognize that for the domain we are interested in $t \in [-k, k]$, $F(t) = 1$. As such, we can replace that term in the integral with 1 to get our new integral:

$$\mathscr{F}(\zeta) = \int_{-k}^{k} e^{-2\pi i \zeta t}\,\mathrm{d}t$$

Apply the $u$-substitution $u = -2\pi i \zeta t \implies \mathrm{d}t = \frac{i}{2\pi\zeta}\,\mathrm{d}u$

$$= \int_{-k}^{k} \frac{i}{2\pi\zeta} e^u \, \mathrm{d}u$$

$$= \frac{i}{2\pi\zeta} \int_{-k}^{k} e^u \, \mathrm{d}u, \qquad \text{(move constant out of integral)}$$

$$= \frac{i}{2\pi\zeta} \left[ e^{-2\pi i \zeta t} \right]_{-k}^{k}$$

To evaluate an integral, calculate $[f(x)]_a^b = f(b) - f(a)$

$$= \frac{i}{2\pi\zeta} \left( e^{-2\pi i \zeta k} - e^{2\pi i \zeta k} \right) \qquad \text{(evaluate integral)}$$

Euler's formula states $e^{\theta i} = \cos\theta + i\sin\theta$, so we can continue evaluating:

$$\mathscr{F}(\zeta) = \frac{i}{2\pi\zeta} \left( [\cos(-2\pi\zeta k) + i\sin(-2\pi\zeta k)] - [\cos(2\pi\zeta k) + i\sin(2\pi\zeta k)] \right)$$

We know that $\cos(-\theta) = \cos(\theta)$ and $\sin(-\theta) = -\sin(\theta)$ since sin is odd and cos is even, so

$$= \frac{i}{2\pi\zeta} \left( \cos(2\pi\zeta k) - i\sin(2\pi\zeta k) - \cos(2\pi\zeta k) - i\sin(2\pi\zeta k) \right)$$

$$= \frac{i}{2\pi\zeta} \left( -2i\sin(2\pi\zeta k) \right)$$

$$= \frac{2\sin(2\pi\zeta k)}{2\pi\zeta} \qquad\qquad (i^2 = -1)$$

$$= \frac{2\sin(2\pi\zeta k)k}{2\pi\zeta k} \qquad\qquad \text{(multiply by } \frac{k}{k}\text{)}$$

$$= 2k\frac{\sin(2\pi\zeta k)}{2\pi\zeta k}$$

but there exists a function $\operatorname{sinc}(x) = \frac{\sin x}{x}$, so we can rewrite

$$= 2k \ \operatorname{sinc}(2\pi\zeta k).$$

We are done! Integrating has brought us to the conclusion that the Fourier transform of a single rectangular pulse is:

$$\mathscr{F}(\zeta) = 2k \ \operatorname{sinc}(2\pi\zeta k)$$

However, the Fourier transform is undefined for $\zeta = 0$ (as we are dividing by 0). We can still use this value by using *limits*. The limit

$$\lim_{\zeta \to 0} \frac{2k\sin(2\pi\zeta k)}{2\pi\zeta k}$$

answers the question "What does [the function] *approach* as $\zeta$ approaches 0?". The answer is not readily apparent as we cannot set $\zeta = 0$. As such, we can utilize the L'Hopital rule, which states that

$$\lim_{x \to a} \frac{f(x)}{g(x)} = \lim_{x \to a} \frac{f'(x)}{g'(x)}$$

(provided the latter limit exists). Thus, by differentiating numerator and denominator, we may be able to arrive at a value:

$$
\begin{aligned}
\lim_{\zeta \to 0} \frac{k \sin(2\pi\zeta k)}{\pi\zeta k} &= \lim_{\zeta \to 0} \frac{\frac{\mathrm{d}}{\mathrm{d}\zeta}(k \sin(2\pi\zeta k))}{\frac{\mathrm{d}}{\mathrm{d}\zeta}(\pi\zeta k)} \\
&= \lim_{\zeta \to 0} \frac{k[\cos(2\pi\zeta k) \cdot 2\pi k]}{\pi k} \qquad \text{(apply chain rule)} \\
&= \lim_{\zeta \to 0} \frac{2\pi k^2 \cos(2\pi\zeta k)}{\pi k} \\
&= \lim_{\zeta \to 0} 2k \cos(2\pi\zeta k)
\end{aligned}
$$

We can now substitute $\zeta = 0$ to evaluate the limit:

$$= 2k \cos(0) = 2k \therefore \mathscr{F}(0) = 2k.$$

This function is plotted for $k = 1$ in Figure 9. From this, we derive that the only defining

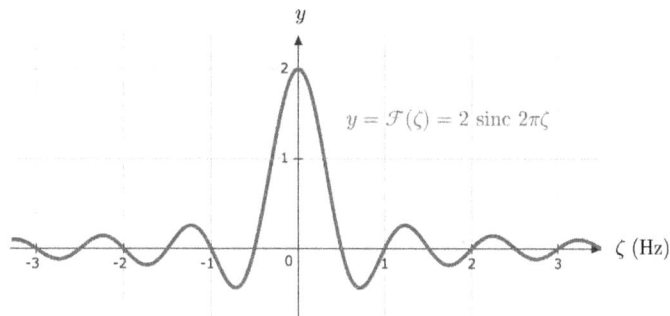

Figure 9: The Fourier transform of the pulse.

factor of the Fourier transform of a square pulse is its width, which is $2k$. The domain of graph in Figure 9 represents frequency in hertz and the corresponding $y$-value the prevalence of the frequency in the original pulse.

## 3.4 In real life: the Discrete Fourier Transform

In its stated form, the FT cannot be used in real life. This is because it requires that the input data be *continuous*—that is, that $f(t)$ be well defined for all $t \in \mathbb{R}$. In real life, we usually sample air pressure at a certain *sample rate* and thus end up with a very finite number of $(t, f(t))$ pairs. As such, the Fourier transform was adapted into a slightly similar algorithm suited for real-life use: the *discrete* Fourier transform, which deals with discrete data (that is, data which is not continuous). Using the framework developed in the above section, we simply adjust the parts which require continuous data. Assume we have a signal with $N$ samples of air pressure, equally spaced in time. $s[n-1]$ is defined to be the $n$th sample of air pressure. Define $\mathscr{S}$ as the discrete Fourier transform of $s$ and as such $\mathscr{S}[\zeta]$ as the complex number representing the frequency $\zeta$. Then, we can define

$$\mathscr{S}[\zeta] = \sum_{n=0}^{N-1} s[n] e^{\frac{-2\pi}{N} i \zeta n} \tag{2}$$

The amendments made to Equation 1 are to replace the integral with a summation and split the unit circle into $N$ discrete parts (through $\frac{2\pi}{N}$) instead of integrating over infinitely many angles. We then calculate $\mathscr{S}[\zeta]$ for as many $\zeta$ as we wish to test.

Every single element in $\mathscr{S}$ represents, in essence, a sinusoidal wave with a different frequency. Every complex number $\mathscr{S}[\zeta]$ represents information about the role of a sinusoidal wave with frequency $\zeta$ in the original harmonic. More specifically, the *modulus* of that complex number tells us the abundance of that frequency and the *argument* of the complex number tells us the extent to which that constituent sinusoidal is shifted.

### 3.4.1 Example of the Discrete Fourier Transform with a simple signal

Suppose we have a recording of a microphone which provided $N = 4$ distinct samples of data as shown in the following table.

| $n$ | 0 | 1 | 2 | 3 |
|---|---|---|---|---|
| $s[n]$ | 1 | 0 | -1 | 0 |

A general understanding is that for $N$ samples, we calculate $N$ different frequency bins.

As such, we will calculate $\mathscr{S}[\zeta]$ for $\zeta \in [1,4] \cap \mathbb{Z}$. We begin by calculating $\mathscr{S}[0]$. By definition in Equation 2, this value is

$$\mathscr{S}[0] = \sum_{n=0}^{3} s[n]e^{\frac{-2\pi}{4}i(0)(n)} = s[0]e^{\frac{-2\pi}{4}i(0)(0)} + s[1]e^{\frac{-2\pi}{4}i(0)(1)} + s[2]e^{\frac{-2\pi}{4}i(0)(2)} + s[3]e^{\frac{-2\pi}{4}i(0)(3)}$$

$$= s[0] + s[1] + s[2] + s[3]$$
$$= 1 + 0 + (-1) + 0 = 0 + 0i$$

which is truly the easiest value to compute as the index of $e$ is reduced to 0 by the $\zeta = 0$ term. Now for the next value, $\zeta = 1$:

$$\mathscr{S}[1] = s[0]e^{\frac{-2\pi}{4}i(1)(0)} + s[1]e^{\frac{-2\pi}{4}i(1)(1)} + s[2]e^{\frac{-2\pi}{4}i(1)(2)} + s[3]e^{\frac{-2\pi}{4}i(1)(3)}$$

but by Euler's formula, $e^{-i\pi} = \cos(-\pi) + i\sin(-\pi) = -1$

$$= (1)e^0 + (0) + (-1)e^{-i\pi} + (0)$$
$$= 2 + 0i$$

We complete the same process for the next candidate frequencies:

$$\mathscr{S}[2] = (1) + (0) + \left(-e^{\frac{-2\pi}{4}4i}\right) + (0) = 0 + 0i$$
$$\mathscr{S}[3] = (1) + (0) + \left(-e^{\frac{-2\pi}{4}6i}\right) + (0) = 2 + 0i$$

It should now be noted that this example is purely to demonstrate the DFT algorithm. Real signals consist of millions of samples and millions of candidate frequencies. It's difficult to pull real insight from this short demonstration. A sampling technicality states that frequencies $\zeta \geq \frac{N}{2}$ may be discarded (this is the Nyquist limit—out-of-scope here), so our final table is:

| $\zeta$ | 0 | 1 | 2 | 3 |
|---|---|---|---|---|
| $\mathscr{S}[\zeta]$ | $0 + 0i$ | $2 + 0i$ | | |

Remember that the complex number returned for each frequency has two pieces of information—the modulus and the argument. As previously stated, the modulus represents the frequency's abundance and the argument the phase shift. For each $\zeta$, let's take the modulus and argument of the complex number:

| $\zeta$ | 0 | 1 | 2 | 3 |
|---|---|---|---|---|
| $\mathscr{S}[\zeta]$ | $0 + 0i$ | $2 + 0i$ | | |
| $\mathrm{Mod}(\mathscr{S}[\zeta])$ | 0 | 2 | | |
| $\mathrm{Arg}(\mathscr{S}[\zeta])$ | 0 | 0 | | |

Real audio samples contain hundreds of thousands of data points and thus require a method in which we graph $\zeta$ versus $\mathrm{Mod}(\mathscr{S}[\zeta])$, for example, and find peaks in the abundance of frequencies. In this case, however, simple inspection reveals that a frequency of $\zeta = 1$ Hz has unusually high modulus (2 versus 0 otherwise). Its corresponding argument is 0, corresponding to absolutely no phase shift. We then pull the insight: "there is a component pure tone of 1 Hz which is not phase shifted". In fact, this is correct—the harmonic used in this example was simply that one pure tone (for simplicity). However simplistic, this preliminary example shows how one can use a table of discrete samples in order to calculate the individual overtone frequencies that make up the sampled signal.

## 4    Real-life acoustic experiment

I set out to apply the discrete Fourier transform in a real-life situation and decompose an artificially created harmonic frequency by recording it with a real microphone device, exporting the raw sound pressure samples to some data format, writing code and implementing my own discrete Fourier transform programmatically, and seeing if I am able to extract individual overtones from such analysis. I knew that this would be an ambitious task, but the idea of implementing an algorithm that has truly stood the test of time and using it in a manner which so resembles professionals excited me. I reached out to my physics teacher so that he could set me up with the proper equipment and assist me in the endeavour.

## 4.1 Experimental setup

The experimental setup consisted of three loudspeakers arranged in a circular pattern with a microphone in the middle. Each loudspeaker was connected to a separate tone generator (so that the speaker would emit that single pure tone). The microphone was connected to a LabQuest interface which communicated with my laptop computer. The experimental setup is illustrated diagrammatically and as a photograph in Figure 10. Initially, data collection was rough. During initial testing with only one speaker active (effectively one pure tone),

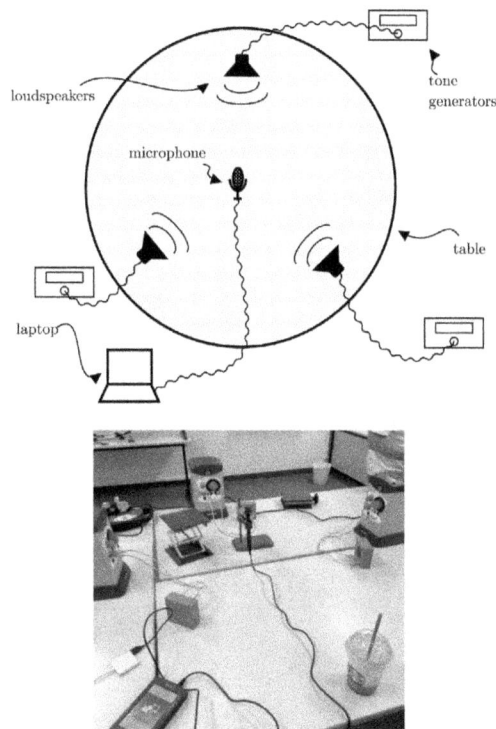

Figure 10: The experimental setup, illustrated both diagrammatically and as a photograph.

found that the microphone was not picking the pure tone up (and instead was returning a "dirty" signal). The next step was to eliminate vibrations from the table by elevating the

microphone using a clamp stand. Elevating the microphone with the clamp stand helped
somewhat, but eventually, the issue was narrowed down to a bad microphone. It was found
that the microphone had a very low sample rate—by today's standards, a microphone is
considered passable if it samples at a rate of 44.1kHz, while this microphone samples at
1kHz. After replacing the microphone with one with a much higher sample rate, testing
showed that the signal that was being picked up was clean. For reasons of simplicity,
however, I settled on using only two loudspeakers of the three. They were set to emit
sounds at frequencies of 711 Hz and 188 Hz. A time series was yielded with air pressure as
the $y$-axis. The time series is shown in Figure 11.

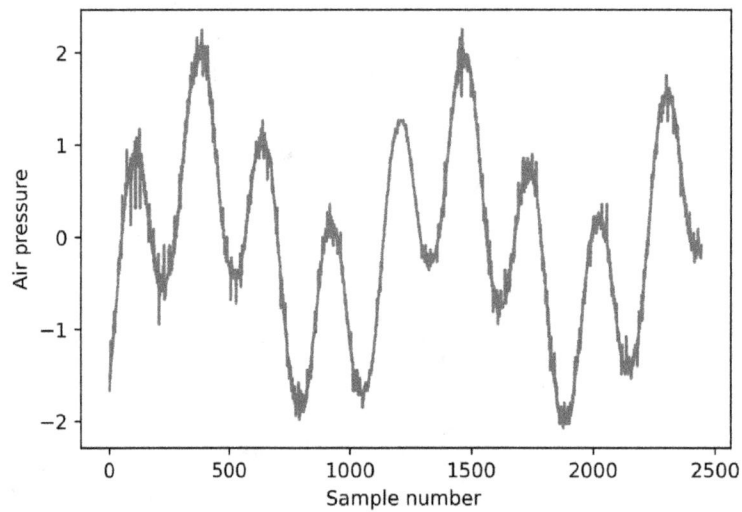

Figure 11: The real data yielded from the microphone. 2500 samples shown in the graph.

The data was easily represented in a tabular manner, a snippet of which is shown below:

| Time | Pressure |
|---|---|
| 0.00000000 | -1.66539500 |
| 0.00512000 | -1.53035100 |
| 0.01024000 | -1.48533600 |
| 0.01536000 | -1.53035100 |
| 0.02048000 | -1.44032100 |
| 0.02560000 | -1.44032100 |
| 9771 more ⋰ | 9771 more ⋰ |

In total, I collected 9771 samples. The next step was to program an algorithm that, given a set of numbers $s$, computed an analogous set of numbers $\mathscr{S}$ that is the discrete Fourier transform of $s$. After a few trials and much head-scratching, I came up with the following (albeit slow) algorithm. The algorithm is detailed in Algorithm 1 and its source code is shown in Appendix 1.

---

**Algorithm 1:** discrete Fourier transform

---

**Result:** discrete fourier transform $\mathscr{S}$
Read sample list $s$;
Let $\zeta = 0$;
Let $N$ be the number of samples;
**for** *the number of samples in s* **do**
    Let $\mathscr{S}[\zeta] = 0$, initially.
    **for** *each sample in s $s[n]$* **do**
        Calculate the value $s[n]e^{\frac{-2\pi}{N}i\zeta n}$ and add it to what we have in $\mathscr{S}[\zeta]$.
    **end**
    Increment $\zeta$ by 1;
**end**

---

This algorithm is largely inefficient, and the number of operations it must complete grows exponentially as does the input size. My input sound only lasted for 50 milliseconds, but the Fourier transform took almost 3 minutes to compute! After obtaining the Fourier transform of the signal $\mathscr{S}[\zeta]$, I plotted the frequency $\zeta$ against the modulus and, separately, the argument of the Fourier transform. The results are shown in Figure 12.

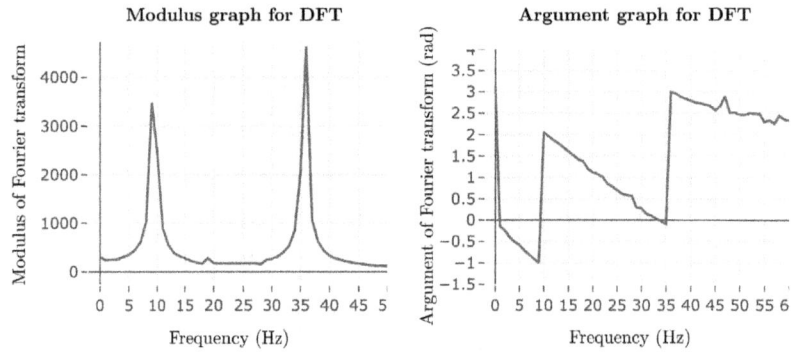

Figure 12: The modulus and argument of the Fourier transform of my real audio signal.

Recall that the original sound was made from two loudspeakers at two different frequencies. It is abundantly clear in the modulus graph that while background frequencies are present, there are two major peaks which bode well with the knowledge that the harmonic contained two pure tones. The argument graph is less helpful in frequency identification, but does roughly tell us the angular position of each of these two component frequencies. I could estimate the frequency bins by eye. However, I wanted this process to be fully automated, and so I passed the DFT modulus graph through a *peak detection algorithm* which automatically identified peaks, as in Figure 13. The algorithm detected peaks at frequency bins 9 and 36.

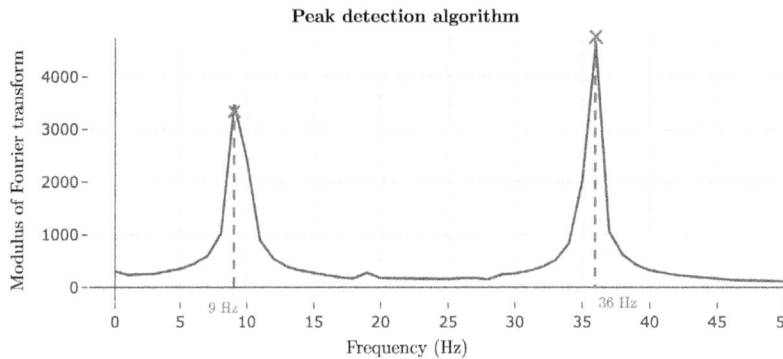

Figure 13: The peak detection algorithm identified peaks at frequency bins 9 and 36.

174

## 4.2 Finishing it up

A final transformation to the data [9, 36] must be done in order to retrieve the true constituent frequencies of the harmonic. Because we were not working in real time values but rather in sample numbers ($n = 3$ did not correspond to the third second but rather the third sample), we must scale the values [9, 36] by the following factor:

$$\frac{R}{N}$$

where $R$ is the sampling rate in hertz and $N$ the number of samples. Given instrument data, this factor was calculated to be 19.991.

$9 \times 19.991 \approx 179$ Hz and $36 \times 19.991 \approx 719$ Hz. Recall that this tone was made from two loudspeakers set at the original frequencies 188 Hz and 711 Hz. These were successfully recovered with an accuracy of 98.8%! This was a momentous achievement for me, using an algorithm I had no experience with and that I learned and implemented from scratch.

# 5 Conclusion

This investigation was a huge success and an amazing learning opportunity for me. I started the investigation by diving into the physics of sound and its periodic nature. Upon realising that most sounds in nature are combinations of pure tones, I set out for a method to "unmix" several mixed frequencies. The investigation led into the Fourier transform, which I built from the ground up using trigonometric concepts as well as infinitesimal calculus. Armed with the formula for a Fourier transform, I practiced on a rectangular pulse and applied L'Hopital's rule to the obtained Fourier transform to evaluate it at $\zeta = 0$. In the ultimate test of understanding of an algorithm that has truly stood the test of time and applies to fields as diverse as audio analysis and quantum physics, I wrote my own implementation of the Discrete Fourier Transform and applied it to real experimental data. This was truly an ambitious project, and it would have been a success even if I had not succeeded in retrieving the source frequencies. When my own implementation of the DFT succeeded in retrieving the frequencies, however, I truly realised the hill I had successfully climbed. I now realise that a single wave in the time domain can carry much more information by encapsulating pure tones inside it. I am now excited to explore the possibility of transmitting data in waves by encoding them as a sum of frequencies (and decoding with a DFT). I am already working on prototype code that may work to this effect. I also hope to investigate the entirety of the square wave (not just one pulse)—this requires knowledge of the Fourier series, which I do not currently possess.

## 6    Bibliography

Cheever, E. (2008). Introduction to the fourier transform. https://lpsa.swarthmore.edu/Fourier /Xforms/FXformIntro.html.

Johns Hopkins University. (2008). The Propagation of sound. https://pages.jh.edu/virtlab/ray/ acoustic.htm.

Weisstein, E. W. (2011). Fourier Transform. from Wolfram MathWorld. https://mathworld.wolfram .com/FourierTransform.html.

Weisstein, E. W. (2011). Discrete Fourier Transform. from Wolfram MathWorld. https://mathworld. wolfram.com/DiscreteFourierTransform.html.

Roberts, S. (2016). Lecture 7 - The Discrete Fourier Transform. Oxford Information Engineering.
https://www.robots.ox.ac.uk/ sjrob/Teaching/SP/l7.pdf.

# 7. MAXIMIZING THE SPINNING TIME OF A TOP

Author: Jennifer A
Moderated Mark: 19/20
Level: Math AA HL

*Section 1 : Introduction*

In the movie *Inception*, a spinning top is used by the protagonist to determine the reality they are in. As an avid fan of the movie, I bought myself a spinning top. The way the top stays upright for an extended period fascinated me. Whilst spinning the top, I noticed the shape of this top were different from the ones encountered in my childhood, which often have a circular base. It made me wonder – how does its shape affect the duration it spins for? What is the most efficient shape to maximize its spinning time ?

Figure 1 – Spinning top from the movie *Inception* [1]

When the opportunity arose, I was determined to satisfy my curiosity. Further background research has made me realised the complexity, and a lack of available information on this matter, making it even more exciting for me to carry out this investigation.

The aim of my exploration is to maximize the spinning time of a top through varying its base shape. To effectively investigate the influence of shape on spinning time, I will be keeping other variables, such as volume, density and initial angular velocity fixed. Moreover, to simplify the problem, I will assume that the torque exerted by air drag is independent of the shape of the top (but in reality, drag force depends on surface area and shape). This is to keep the influence of friction and air drag consistent, allowing me to solely focus on angular momentum. With the following simplifications in mind, I will first determine factors that affect the spinning time of the top.

## Section 2 : Background Physics

### 2.1 Forces on a Spinning Top

To understand the motions of a spinning
object, I must first understand torque $\vec{\tau}$ and
angular momentum, $\vec{L}$. Torque, defined as
$\vec{\tau} = \vec{r} \times \vec{F}$ (vectors quantities will be
explained in *section 2.2*), is the result of
applying a force, $\vec{F}$ to rotate an object around
an axis, where $\vec{r}$ is the distance from the
axis of rotation to the object (Figure 2 ) [2].

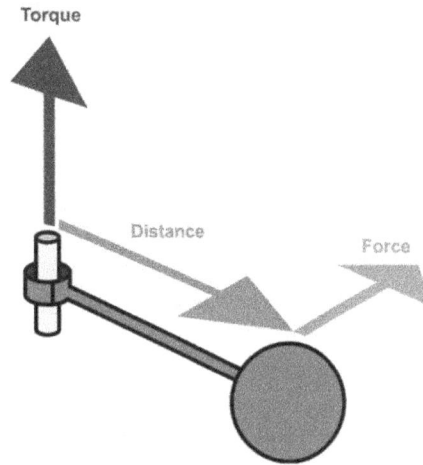

Figure 2– Forces exerted on a spinning top [2]

On the other hand, angular momentum is the rotational analog of linear momentum $\vec{p}$, defined
as $\vec{L} = \vec{r} \times \vec{p}$. There are 2 special features to angular momentum : it is a conserved quantity,
and that its rate of change is torque, $\vec{\tau} = \frac{d\vec{L}}{dt}$ [3]. This will be explored and discussed in more
detail in *section 3.1*.


To determine factors influencing the spinning time of a top, I first examined the forces acting
on it. Figure 3 shows an ideal cone-shaped top rotating about its axis of symmetry on a flat
surface.

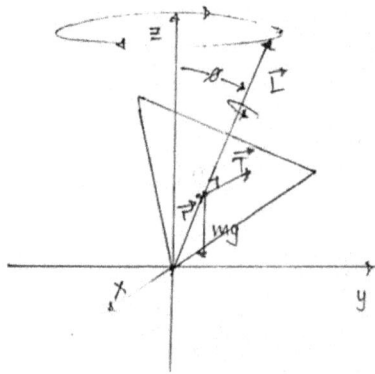

Figure 3 – Diagram of a spinning top    Figure 4 – Forces exerted on a spinning top

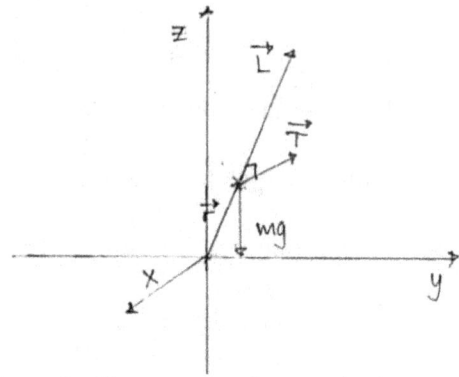

When a top is spinning, it has got an angular momentum $\vec{L}$ along the axis of the top. The force of gravity $m\vec{g}$ , where $m$ is the mass of the top and $\vec{g}$ is the gravitational acceleration, creates a torque $\vec{\tau}$ perpendicular to $\vec{L}$, which in turn changes the direction of the top's rotational axis (Figure 4).

When gravity is the only external force acting on the top, the magnitude of angular momentum is constant but it's direction changes. Hence the spinning top precesses in a constant precession angle $\phi$ provided that the angular speed is above a certain threshold. Below this threshold, the top starts to wobble and precession of the top becomes complicated (which will be discussed in *section 5.2*) [4]. Such complications will be ignored in my exploration.

However, in reality, air drag and friction from the ground creates an external torque that causes a deceleration. This slows the top down and reduces the magnitude of angular momentum. When the angular momentum drops below a certain threshold, the top no longer precesses steadily and gravity causes the top to eventually fall over.

To maximize the spinning time of a top, I would have to minimize the deceleration caused by air drag and friction. By assuming that the torque exerted by air drag is independent of the shape of the top (stated in *section 1*), this allows me to solely focus on the change of angular momentum. To do so, I first looked into the relationship between angular momentum and torque of a point mass.

## 2.2 Dot and Cross Product

This investigation will be revolved around vector quantities, hence it is important to understand how to manipulate vectors. As opposed to scalar quantities that only have a magnitude, vector quantities have both magnitude and direction. Vectors are denoted $\vec{a}$ and can be multiplied using dot product (e.g. $\vec{a} \cdot \vec{b}$) or cross product (e.g. $\vec{a} \times \vec{b}$).

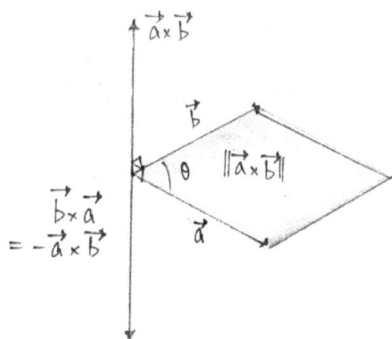

Scaler product/ dot product

$$\vec{a} \cdot \vec{b} = \|\vec{a}\|\|\vec{b}\|\cos\theta$$

Vector product/ cross product

$$\vec{a} \times \vec{b} = \|\vec{a}\|\|\vec{b}\|\sin\theta$$

where the magnitude of vector $\vec{a}$ is denoted by $\|\vec{a}\|$ [5]

Figure 5 – Cross Product

The resultant vector from the cross product has a direction that is perpendicular to both $\vec{a}$ and $\vec{b}$, and has a magnitude equal to the area of a parallelogram with vectors $\vec{a}$ and $\vec{b}$ as sides [6]. The direction of the resultant vector can be determined by the right hand rule. It is important to note that the order in which the vector is written in determines the direction of the resultant vector, where $\vec{a} \times \vec{b} = -\vec{b} \times \vec{a}$ (Figure 5)

181

*3.1 Torque and Angular Momentum of a Point Mass*

The angular momentum of an extended object is defined as the integral sum of its constituent particles [7]. That is, by adding up the angular momentum of particles that make up an object, I can find its angular momentum. To do so, I first considered the case of a point particle.

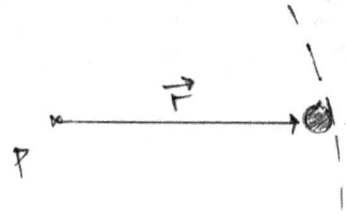

Figure 6 – Diagram of Point Particle

Let $\vec{r}$ be the displacement of the particle from a point $P$ and $\vec{p} = m\vec{v}$ be the linear momentum of the particle, where $m$ is the mass of the particle and $\vec{v}$ is its linear velocity (Figure 6). As angular momentum is defined as $\vec{L} = \vec{r} \times \vec{p}$ (*section 2.1*) then the angular momentum about point $P$ is would be

$$\vec{L} = \vec{r} \times \vec{p} = m\vec{r} \times \vec{v} \quad [8]$$

It is important to note that the magnitude of $\vec{L}$ depends on the choice of reference point $P$.

If there is a force $\vec{F}$ acting on the particle, then according to Newton's second law $\vec{F} = \frac{d\vec{p}}{dt}$, there will be a change in momentum, $\vec{p}$. This would result in a torque, $\vec{\tau}$

$$\vec{F} = \frac{d\vec{p}}{dt}$$

$$\rightarrow \vec{r} \times \vec{F} = \vec{r} \times \frac{d\vec{p}}{dt}$$

$$\vec{\tau} = \vec{r} \times \frac{d\vec{p}}{dt}$$

As angular momentum, $\vec{L} = \vec{r} \times \vec{p}$, I can write torque in terms of angular momentum

To do so, I used the product rule for cross product, $\frac{d(\vec{r} \times \vec{p})}{dt} = \vec{r} \times \frac{d\vec{p}}{dt} + \frac{d\vec{r}}{dt} \times \vec{p}$ [9]. Thus

$$\vec{\tau} = \vec{r} \times \frac{d\vec{p}}{dt} = \frac{d(\vec{r} \times \vec{p})}{dt} - \frac{d\vec{r}}{dt} \times \vec{p} = \frac{d(\vec{r} \times \vec{p})}{dt} - \vec{v} \times (m\vec{v}) = \frac{d\vec{L}}{dt}$$

182

Here, I have deduced that torque is proportional to the rate of change of angular momentum. This shows that without external torque, angular momentum is conserved. However, understanding this relationship alone is not sufficient to determine how do I minimize deceleration by external torques (such as air drag and friction). Thus, my next step will be establishing the relationship between torque, angular momentum and angular acceleration.

*3.2 From Angular Momentum to Angular Acceleration*

If the particle is undergoing any sort of circular motion (doesn't necessarily have to be uniform), then I can rewrite linear velocity $\vec{v}$ in terms of angular velocity $\vec{\omega}$, defined via $\vec{v} = \vec{\omega} \times \vec{r}$

$$\vec{L} = m\vec{r} \times (\vec{\omega} \times \vec{r}) = mr^2 \vec{\omega}$$

From this, I have established that $\vec{L} \propto \vec{\omega}$. The proportionality constant is called the *moment of inertia* of the particle with respect to the point $P$, which is conventionally denoted by $I$, thus $\vec{L} = I\vec{\omega}$.

In the case of a point particle, $I = m\vec{r}^2$. The moment of inertia will differ depending on the mass distribution of the object, but $\vec{L} \propto \vec{\omega}$ is general [11]. With this, I can rewrite torque acting on the point particle in terms of its moment of inertia.

$$\vec{\tau} = \frac{d\vec{L}}{dt} = mr^2 \frac{d\vec{\omega}}{dt} = I \frac{d\vec{\omega}}{dt}$$

To simplify the relationship, I rewrote angular velocity in terms of angular acceleration, $\vec{\alpha}$, where $\vec{\alpha} = \frac{d\vec{\omega}}{dt}$. Hence

$$\vec{\tau} = \frac{d\vec{L}}{dt} = I \frac{d\vec{\omega}}{dt} = I\vec{\alpha}$$

The relationship $\vec{\tau} = I\vec{\alpha}$ is generally true provided that the moment of inertia remains constant [12]. From $\vec{\tau} = I\vec{\alpha}$, I realised that the greater the moment of inertia, the smaller the angular deceleration caused by air drag and friction. This is based on the earlier assumption that torque exerted by air drag is independent of the shape of the top. By assuming this, the product of $I\vec{\alpha}$ remains a fixed value. Thus, the spinning time of a top will be maximized when the moment of inertia is maximized.

*3.3 Moment of Inertia of a Circular Disc*

As mentioned in *section 3.1*, the angular momentum of an extended object is defined as the integral sum of its constituent particles. Due to keeping the initial angular velocity constant, the angular momentum of a spinning top is directly proportional to its moment of inertia ($\vec{L} = I\vec{\omega}$). Hence I can calculate the moment of inertia of an object by adding up the moment of inertia of its constituent particles. To simplify calculations, I will consider a spinning top as a stack of circular discs, where its moment of inertia will be the sum of the moment of inertia of the discs.

Suppose there is a circular disk with area $A$, thickness $\Delta h$, radius $R$ and mass $m$. To calculate the moment of inertia of the disc, I broke it down into $n$ small fragments and treated each fragment as point particles (Figure 7). For the $i$th fragment, its moment of inertia would be

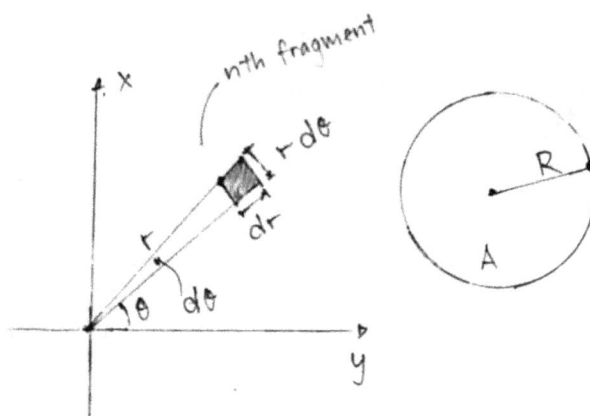

Figure 7 – Diagram of Fragment and Disc

$$\Delta I_i \cong r_i^2 \Delta m_i$$

Where $r_i$ is the distance from the axis of rotation to the $i$th fragment. Hence the sum of all particles would be

$$I_d = \sum_{i=1}^{n} \Delta I_i \cong \sum_{i=1}^{n} r_i^2 \Delta m_i$$

Where $I_d$ is the moment of inertia of a disc. At this point, I encountered complications in integrating $\Delta m_i$. This is because $\Delta m_i$ is dependent upon the location of the fragment (which makes it a function of $r_i$. To proceed further, I rewrote $\Delta m_i$ in terms of $r_i$. To do so, I first defined the density of the material as $\rho = \frac{m}{V} = \frac{m}{Ah}$ where $V$ is its volume [ 13]. Rearranging and subsituting for $\Delta m_i$

$$\Delta m_i = \rho \Delta V_i = \rho\, \Delta h\, \Delta A_i = \rho \Delta h \cdot \Delta r \cdot r_i \Delta \theta$$

Then from Figure 7, I realised that the area of a fragment, $\Delta A_i$ can be obtained through multiplying its sides, thus $\Delta A_i = \Delta r \cdot r_i \Delta \theta$. With this, I can rewrite the equation as

$$\Delta m_i = \rho\, \Delta h\, \Delta A_i = \rho \Delta h \cdot \Delta r \cdot r_i \Delta \theta$$

Subsituting this into the formula for moment of inertia

$$I_d \cong \sum_{i=1}^{n} r_i^2 \cdot \rho \Delta h \cdot \Delta r \cdot r_i \Delta \theta$$

By taking the limit $n \to \infty$, the summation becomes an integral

$$I_d = \int_{disc} r^3 \cdot \rho \Delta h \cdot dr\, d\theta$$

where $\int_{disc}$ is an abbreviation for integration over the whole disc. As the disc is circular with a radius of $R$, the $r$-intergral is from 0 to $R$ and the $\theta$ integral is from 0 to $2\pi$. Therefore the integral would be

$$\int_0^{2\pi} \left( \int_0^R r^3 \cdot \rho \Delta h \cdot dr \right) d\theta$$

Taking out the constant, $\rho \Delta h$

$$= \rho \Delta h \int_0^{2\pi} \left( \int_0^R r^3 \, dr \right) d\theta$$

Integrating and substituting the boundaries of $\int_0^R r^3 \, dr$

$$= \rho \Delta h \int_0^{2\pi} \left[ \frac{r^4}{4} \right]_0^R d\theta$$

$$= \rho \Delta h \int_0^{2\pi} \frac{R^4}{4} \, d\theta$$

Finally integrating and subsituting the boundaries of $\int_0^{2\pi} \frac{R^4}{4} \, d\theta$

$$= \frac{1}{4} \rho R^4 \Delta h [\theta]_0^{2\pi}$$

$$= \frac{1}{4} \rho R^4 \Delta h (2\pi)$$

$$I_d = \frac{1}{2} \rho \pi R^4 \Delta h$$

With the moment of inertia of a single disc, I can then proceed further by using this to calculate the moment of inertia of a top.

186

As mentioned earlier, a cylindrically symmetric body, such as a top, can be considered as stacked discs, where its moment of inertia will be the sum of the moment of inertia of the discs.

Figure 8 shows such a disc with height $H$, spilt into $N$ discs where each of them has a thickness $\Delta h$. The $i$-th disc at height $h_i$ has a radius (as a function of height) $r_i = r(h_i)$. From what I have established above, the moment of inertia for one disc, $\Delta I_i$ would then be

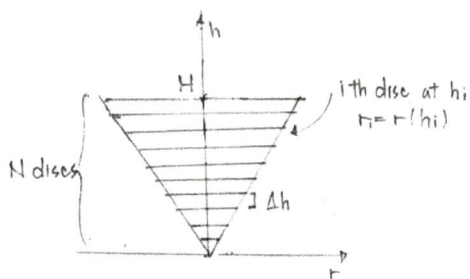

Figure 8 – Stacked Discs

$$\Delta I_i = \frac{1}{2}\rho\pi r_i^4 \Delta h$$

Subsituding $r_i = r(h_i)$

$$\Delta I_i = \frac{1}{2}\rho\pi r_i^4 \Delta h = \frac{1}{2}\rho\pi \cdot r(h_i)^4 \Delta h$$

Hence the moment of inertia of the top, as the sum of the moment of inertia of its respective discs would be

$$I_{top} = \sum_{i=1}^{N} \Delta I_i = \sum_{i=1}^{N} \frac{1}{2}\rho\pi \cdot r(h_i)^4 \, \Delta h$$

Taking the limit $N \to \infty$

$$I_{top} = \frac{1}{2}\rho\pi \int_{0}^{H} r(h)^4 \, dh$$

Where the $h$-integral is from 0 to $H$ due to that being the height of the top.

*4.1 Varying the Shape of a Top*

Now that I have the formula for moment of inertia of a spinning top, I wanted to investigate the effect changing the base shape has on its moment of inertia. To do so, I set the slope of the top as a function of $h$ and investigated the moment of inertia for various shapes.

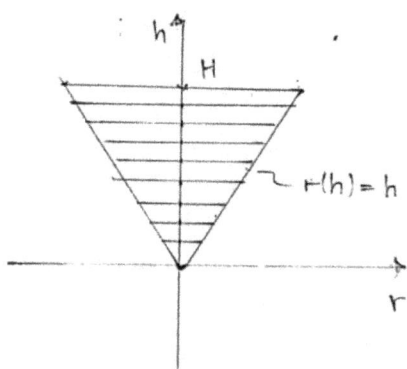

I started off by first considering the most basic top shape – a cone (Figure 9). By substituting $r(h) = h$ and integrating the function, the moment of inertia becomes

$$I = \frac{1}{2}\rho\pi \int_0^H h^4 \, dh = \frac{1}{2}\rho\pi \left[\frac{1}{5}h^5\right]_0^H = \frac{1}{10}\rho\pi H^5$$

Figure 9 – top with sides $r(h) = h$

Substituting the volume of a cone, $V = \frac{1}{3}\pi H^3$ into the moment of inertia of the top

$$I = \frac{1}{10}\rho\pi H^5 = \frac{3}{10}\rho V H^2$$

And from $m = \rho V$

$$I = \frac{3}{10}\rho V H^2 = \frac{3}{10}m H^2$$

Afterwards, I investigated the moment of inertia of tops with sides of various power coefficient of $h$. To enable a better comparison between the moment of inertia of various shapes, I fixed the volume of the tops as V=$\frac{\pi H^3}{3}$ (could be any consistent value, this is just for convenience).

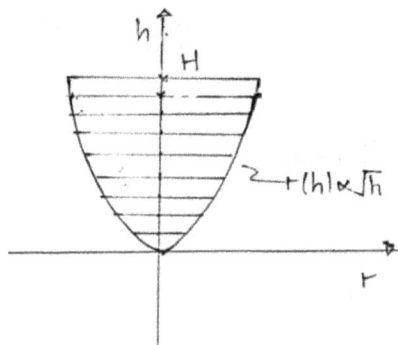

Figure 10 – top with sides $r(h) \propto \sqrt{h}$

The volume of a top with sides $r(h) = k\sqrt{h}$ (Figure 10) can be written as the integral sum of its respective discs. As such a disc would have a volume of $\pi r(h_i)^2$, the volume of such a top would be

$$V \cong \sum_{i=1}^{N} \pi r(h_i)^2 \, \Delta h$$

Taking the limit $N \to \infty$

$$V = \int_0^H \pi r(h)^2 dh$$

The proportionality constant $k$ of a top with volume $V$ and sides $r(h) = k\sqrt{h}$ (Figure 10) could be obtained through substituting $r(h) = k\sqrt{h}$ into its volume, where $V = \frac{\pi H^3}{3}$

$$\frac{\pi H^3}{3} = \int_0^H \pi (k\sqrt{h})^2 \, dh$$

Taking out the constant, $\pi k^2$

$$\frac{\pi H^3}{3} = \pi k^2 \int_0^H (\sqrt{h})^2 \, dh$$

Integrating and substituting the boundaries of $\int_0^H (\sqrt{h})^2 \, dh$

$$\frac{\pi H^3}{3} = \pi k^2 \left[ \frac{1}{2} h^2 \right]_0^H$$

$$\frac{\pi H^3}{3} = \pi k^2 \frac{1}{2} H^2$$

Solving for $k$

$$k^2 = \frac{2}{3} H$$

189

$$k = \sqrt{\frac{2H}{3}}$$

Substituting into $r(h)$

$$r(h) = \sqrt{\frac{2Hh}{3}}$$

With this, I can subsitute $r(h) = \sqrt{\frac{2Hh}{3}}$ into the formula to obtain its moment of inertia

$$I = \frac{1}{2}\rho\pi \int_0^H \left(\sqrt{\frac{2Hh}{3}}\right)^4 dh$$

$$= \frac{1}{2}\rho\pi \int_0^H \frac{4}{9}h^2 H^2 \, dh$$

$$= \frac{4}{18}\rho\pi H^2 \int_0^H h^2 \, dh$$

Integrating and substituting

$$= \frac{4}{18}\rho\pi H^2 \left[\frac{1}{3}h^3\right]_0^H$$

$$= \frac{4}{18}\rho\pi H^2 \cdot \frac{1}{3}H^3$$

Substituting V=$\frac{\pi H^3}{3}$ and from $m = \rho V$

$$I = \frac{2}{9}\rho V H^2 = \frac{2}{9}m H^2$$

From this, I realised that the moment of inertia of a top with sides $r(h) \propto \sqrt{h}$ is smaller than that of a cone-shaped top with sides $r(h) = h$. Thus, I hypothesized that the sharper the base of the top, the larger its moment of inertia.

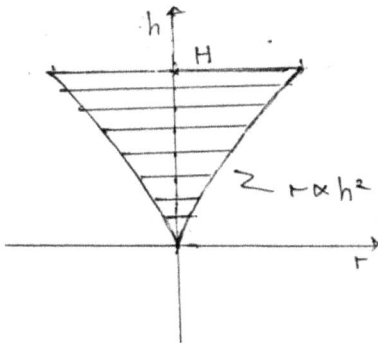

Figure 11 – top with sides $r(h) \propto h^2$

To investigate this, I considered the moment of inertia of a top with sides $r(h) \propto h^2$, which has a sharper base than the 2 cases I have examined above (Figure 11). With $V = \frac{\pi H^3}{3}$, the proportionality constant $k$ would be

$$V = \frac{\pi H^3}{3} = \int_0^H \pi \cdot k^2 h^4 \, dh$$

$$k = \sqrt{\frac{5}{3}} \frac{1}{H}$$

Hence its moment of inertia would be

$$I = \frac{1}{2}\rho\pi \int_0^H \left(kh^2\right)^4 dh = k^4 \cdot \frac{1}{18}\rho\pi \cdot H^9 = \frac{25}{54}\rho\pi H^5 = \frac{25}{54}mH^2$$

As the moment of inertia of a top with sides $r(h) \propto h^2$ appears to have the highest moment of inertia among the cases I have examined so far, this supports my hypothesis that the sharper the base of the top, the larger its moment of inertia.

## 4.2 Optimizing the Shape of a Top

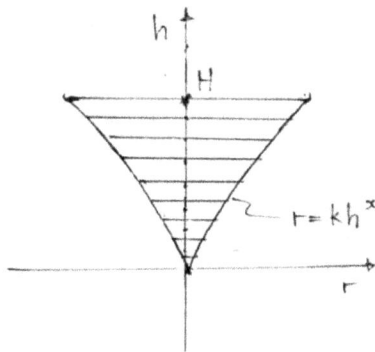

Figure 12 – top with sides $r(h) \propto h^x$

From the cases I examined so far, it appears to be a general trend where the larger the power coefficient of $h$, the larger its moment of inertia. This raises a couple of questions : is there a power coefficient of $h$ in which the moment of inertia is maximized ? If so, what is it ? To find out, I considered a top with sides $r(h) = kh^x$ (Figure 12) and volume $V = \frac{\pi H^3}{3}$.

The proportionality constant $k$ of such a top would be

$$V = \frac{\pi H^3}{3} = \int_0^H \pi(kh^x)^2 \, dh$$

Taking out the constant, $\pi k^2$ and simplifying

$$\frac{H^3}{3} = k^2 \int_0^H h^{2x} \, dh$$

Integrating and substituting the boundaries of $\int_0^H h^{2x} \, dh$

$$\frac{H^3}{3} = k^2 \left[ \frac{1}{2x+1} h^{2x+1} \right]_0^H$$

$$\frac{H^3}{3} = \frac{k^2}{2x+1} H^{2x+1}$$

Rearranging and simplifying

$$\frac{k^2}{2x+1} H^{2(x-1)} = \frac{1}{3}$$

Solving for $k$

$$k^2 = \frac{2x+1}{3H^{2(x-1)}}$$

$$k(x) = \pm \sqrt{\frac{2x+1}{3}} \frac{1}{H^{x-1}}$$

As the top is cylindrically symmetric, it doesn't matter whether I take the positive or negative value of $k(x)$. Still, I will take the positive value of $k(x)$ so that $r$ is positive.

Substituting $k(x)$ into formula for the moment of inertia, $I = \frac{1}{2}\rho\pi \int_0^H r(h)^4 \, dh$ and integrating with respect to $dh$

$$I(x) = \frac{1}{2}\rho\pi \int_0^H (k(x)\, h^x)^4 \, dh$$

$$= \frac{1}{2}\rho\pi \int_0^H k(x)^4 \, h^{4x} \, dh$$

$$I(x) = \frac{1}{2}\rho\pi \cdot k(x)^4 \left[\frac{1}{4x+1} H^{4x+1}\right]$$

This gives me the moment of inertia for a a top with sides $r(h) = kh^x$ and volume $V = \frac{\pi H^3}{3}$ .

To find the value of $x$ in which the moment of inertia is maximized, I will take the derivative of $I(x)$ and set it to zero, calculating $\frac{dI(x)}{dx} = 0$ . Afterwards, I will plot the function to determine whether it is a maximum, minimum or point of inflexion.

$$\frac{dI(x)}{dx} = \frac{1}{2}\rho\pi \cdot \frac{d}{dx}\left(k(x)^4 \left[\frac{1}{4x+1} H^{4x+1}\right]\right) = 0$$

Although the derivative seems complicated at first glance, it can be tackled through a series of chain, product and quotient rules. Let $u_I$ be $k(x)^4$ and $v_I$ be $\frac{1}{4x+1} H^{4x+1}$, thus

$$\frac{d}{dx}\left(k(x)^4 \left[\frac{1}{4x+1} H^{4x+1}\right]\right) = \frac{d}{dx}u_I v_I = u_I\frac{dv_I}{dx} + v_I\frac{du_I}{dx}$$

First, I used chain rule to find $\frac{du_I}{dx}$

$$\frac{du_I}{dx} = \frac{d}{dx}(k(x)^4) = 4k(x)^3\frac{dk(x)}{dx}$$

Afterwards, I used product rule to find $\frac{dv_I}{dx}$

$$\frac{dv_I}{dx} = \frac{d}{dx}\left(\frac{1}{4x+1} H^{4x+1}\right) = -\frac{4}{4x+1} \cdot H^{4x+1} \cdot \ln H - \frac{4}{(4x+1)^2} \cdot H^{4x+1}$$

Here, I used a series of chain rules to find $\frac{d}{dx}\left(\frac{1}{4x+1}\right)$ and $\frac{d}{dx}(H^{4x+1})$

Subsituting $\frac{du_I}{dx}$ and $\frac{dv_I}{dx}$ into $\frac{d}{dx}u_I v_I$

$$\frac{d}{dx}u_I v_I = u_I\frac{dv_I}{dx} + v_I\frac{du_I}{dx}$$

$$= k(x)^4 \cdot \left[-\frac{4}{4x+1} \cdot H^{4x+1} \cdot \ln H - \frac{4}{(4x+1)^2} \cdot H^{4x+1}\right] + 4k(x)^3\frac{dk(x)}{dx} \cdot \left[\frac{1}{4x+1} H^{4x+1}\right]$$

Rearranging and simplifying

$$\frac{d}{dx}\left(k(x)^4\left[\frac{1}{4x+1}H^{4x+1}\right]\right) = k(x)^3\frac{4}{4x+1}H^{4x+1}\left[k(x)\left(\ln H - \frac{1}{4x+1}\right) + \frac{dk(x)}{dx}\right]$$

Subsituting back into $\frac{dI(x)}{dx}$

$$\frac{dI(x)}{dx} = \frac{1}{2}\rho\pi \cdot k(x)^3\frac{4}{4x+1}H^{4x+1}\left[k(x)\left(\ln H - \frac{1}{4x+1}\right) + \frac{dk(x)}{dx}\right]$$

Solving for $\frac{dI(x)}{dx} = 0$

$$\frac{1}{2}\rho\pi \cdot k(x)^3\frac{4}{4x+1}H^{4x+1}\left[\left(\ln(H) - \frac{1}{4x+1}\right)k(x) + \frac{dk(x)}{d(x)}\right] = 0$$

To proceed further from this point, I found the derivative of $k(x)$ via $\frac{dk(x)}{dx} = \frac{d}{dx}k(x)$.

Subtituding $k(x) = \sqrt{\frac{2x+1}{3}}\frac{1}{H^{x-1}}$

$$\frac{dk(x)}{d(x)} = \frac{d}{dx}\left(\sqrt{\frac{2x+1}{3}}\right)\frac{1}{H^{x-1}}$$

Here, I used the product rule. Let $u_k$ be $\sqrt{\frac{2x+1}{3}}$ and $v_k$ be $\frac{1}{H^{x-1}}$, thus

$$\frac{d}{dx}\left(\sqrt{\frac{2x+1}{3}}\right)\frac{1}{H^{x-1}} = \frac{d}{dx}u_k v_k = u_k\frac{dv_k}{dx} + v_k\frac{du_k}{dx}$$

Using chain rule,

$$\frac{du_k}{dx} = \frac{d}{dx}\left(\sqrt{\frac{2x+1}{3}}\right) = \frac{3}{4\left(\frac{2x+1}{3}\right)^{\frac{1}{2}}} = \frac{1}{3}\left(\frac{2x+1}{3}\right)^{-\frac{1}{2}}$$

Using quotient rule,

$$\frac{dv_k}{dx} = \frac{d}{dx}\left(\frac{1}{H^{x-1}}\right) = -H^{-x+1}\cdot\ln H$$

Subsituting $\frac{du_k}{dx}$ and $\frac{dv_k}{dx}$ into $\frac{d}{dx}u_k v_k$

194

$$\frac{d}{dx} u_k v_k = u_k \frac{dv_k}{dx} + v_k \frac{du_k}{dx}$$

$$= -\left(\frac{2x+1}{3}\right)^{\frac{1}{2}} \cdot \frac{1}{H^{x-1}} \ln H + \frac{1}{3}\left(\frac{2x+1}{3}\right)^{-\frac{1}{2}} \cdot \frac{1}{H^{x-1}}$$

Rearranging and simplifying

$$= \frac{1}{H^{x-1}}\left(\frac{2x+1}{3}\right)^{\frac{1}{2}}\left(\frac{1}{2x+1} - \ln H\right)$$

Subtituding $k(x) = \sqrt{\frac{2x+1}{3}} \, \frac{1}{H^{x-1}}$

$$\frac{dk(x)}{d(x)} = k(x)\left[\frac{1}{2x+1} - \ln H\right]$$

Subsituding $\frac{dk(x)}{d(x)}$ back into $\frac{dI(x)}{d(x)}$

$$\frac{dI(x)}{d(x)} = \frac{1}{2}\rho\pi \cdot k(x)^3 \frac{4}{4x+1} H^{4x+1}\left[\left(\ln H - \frac{1}{4x+1}\right)k(x) + \left(\frac{1}{2x+1} - \ln H\right)k(x)\right] = 0$$

Taking out the constant, $k(x)$

$$\frac{1}{2}\rho\pi \cdot k(x)^3 \frac{4}{4x+1} H^{4x+1}k(x)\left[\left(\ln H - \frac{1}{4x+1}\right) + \left(\frac{1}{2x+1} - \ln H\right)\right] = 0$$

Dividing both sides by $\frac{1}{2}\rho\pi \cdot k(x)^3 \frac{4}{4x+1} H^{4x+1}k(x)$

$$\ln H - \frac{1}{4x+1} + \frac{1}{2x+1} - \ln H = 0$$

$$= \frac{2x}{(4x+1)(2x+1)} = 0$$

$$x = 0$$

However $\frac{dI(x)}{dx} = 0$ doesn't indicate the nature of the function – this could be a maximum, minimum or point of inflexion. As calculating the $2^{nd}$ derivative would be unefficient, I decided to plot the anti-derivative of $\frac{dI(x)}{dx}$, which is $I(x)$, on a graph. This will enable me to examine the nature of this point.

$$I(x) = \frac{1}{2}\rho\pi \cdot k(x)^4 \left[ \frac{1}{4x+1} H^{4x+1} \right]$$

Subsituding $k(x) = \sqrt{\frac{2x+1}{3} \frac{1}{H^{x-1}}}$

$$I(x) = \frac{1}{2}\rho\pi \cdot \left[ \frac{(2x+1)^2}{9(4x+1)} \right] \cdot H^5$$

In order to ensure that the function does not have a dimension, I divided both sides by $\rho H^5$

Plotting $\frac{I(x)}{\rho H^5} = \frac{1}{2}\pi \cdot \left[ \frac{(2x+1)^2}{9(4x+1)} \right]$

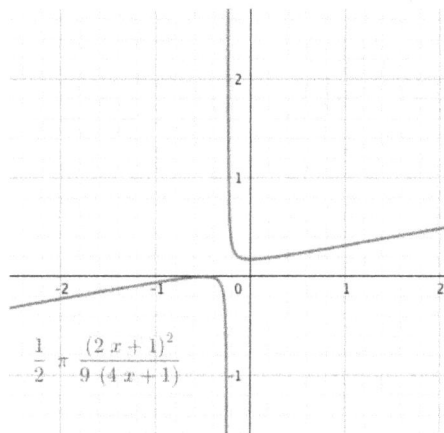

Figure 13 – Graph of $\frac{I(x)}{\rho H^5}$

The graph shows that when $x = 0$, $\frac{dI(x)}{dx}$ is a minimum. As $I < 0$ is not physically possible, values of $I < 0$ will be ignored. From its derivative, I know that there is no maximum, hence as $x$ tends to infinity, moment of inertia $I$ tends to infinity.

This is in accordance with my hypothesis, where I proposed that the sharper the base of the top, the larger its moment of inertia. However, it is physically impractical to have a top with sides $r(h) \propto h^x$ where $x$ is a really large number (demonstrated in Figure 14). This is because such a top would have easily wobble over and could not maintain a steady state due to the fundamental laws of physics.

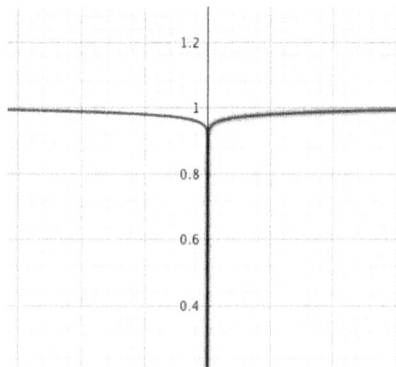

Figure 14 – Graph of $r(h) = h^{80}$

For instance, the top would easily come into contact with the ground when precession occurs, which immediate stops the top from spinning. A possible alternative solution and area of further investigation would be to impose physical constraints to the shape of the top (Figure 16). By limiting the angle between the base of the top and its maximum radius for example, this could decrease the chances of the top falling over due to the sides being in contact with the ground. M

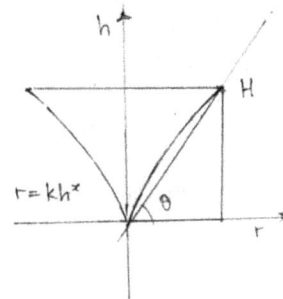

Figure 16 – Possible physical constrains to the shape of the top

197

## Section 5 : Conclusion

### 5.1 Conclusion

The aim of my exploration is to maximize the spinning time of a top through varying its base shape. To do so, I first analyzed forces acting on a spinning top. This has enabled me to identify factors that affect the spinning time of the top. Through my analysis, I realised that to maximize the spinning time of a top, I would have to minimize the deceleration caused by air drag and friction. To do so, I first looked into the relationship between angular momentum and torque of a point mass. Then, I used this knowledge to derive the relationship between angular acceleration and angular momentum. With this, I established that $\vec{L} = I\vec{\omega}$. As initial angular velocity is kept constant in my exploration, the spinning time of a top will be maximized when the moment of inertia is maximized. This is because the larger the moment of inertia, the smaller the angular deceleration caused by air drag and friction thus the longer the duration span.

With the new focus of my exploration on maximizing the moment of inertia of a top through varying its base shape, I looked into how could I calculate the moment of inertia of a spinning top. To do so, I first started by considering the case of a point particle, before moving on to circular discs. By considering a spinning top as a stack of circular discs, I have successfully derived an equation for the moment of inertia of a top based on its slope.

Afterwards, I applied my equation to tops of various shapes, including a cone-shaped top, a top with a circular base and a top with a thin base. From my calculations, I have realised that the top with a thin base had the largest moment of inertia. Thus, I hypothesized that the sharper the base of the top, the larger its moment of inertia.

To test my hypothesis, I considered a top with sides $r(h) = kh^x$ and volume $V = \frac{\pi H^3}{3}$. To

find the value of $x$ in which the moment of inertia is maximized, I calculated the value of $x$ in

which $\frac{dI(x)}{dx} = 0$. Through a complex series of algebraic manipulations, I arrived at a

solution. However $\frac{dI(x)}{d(x)} = 0$ doesn't indicate the nature of the function – this could be a

maximum, minimum or point of inflexion. As calculating the 2nd derivative would be

unefficient, I plotted the anti-derivative of $\frac{dI(x)}{d(x)}$ (which is $I(x)$) on a graph to examine the

nature of this point. My graph of $I(x)$ reveals that there is no maximum to the function, thus

as $x$ tends to infinity, moment of inertia $I$ tends to infinity.


Relating this to the aim of my exploration, I realised that it is physically impartical to have a

top with sides $r(h) \propto h^x$ where $x$ is a really large number. This is due to a number of

physical limitations, outlined in *section 4.2*. However, additional criterias could be imposed

to create a spinning top with the maximum moment of inertia. Such areas of further

investigations will be explored in *section 5.3*.


## 5.2 Evaluation of Findings : Assumptions and Limitations

There are multiple assumptions that I made, which would have affected the validity of my

findings. First, I assumed the torque exerted by air drag is independent of the shape of the

top. In reality however, drag force depends on both surface area and shape of the top. This

was done to simplify the problem, which enabled me to focus on how the shape of the top

affects its moment of inertia, hence time span. Moreover, I made an implicit assumption of

the tops having the same maximum height $H$ when comparing between tops of different

shapes. This was done to allow effective comparison, but if time allows, the influence of

varying the height $H$ of the top should be investigated. Furthermore, I only invested tops with slopes that obey the function, $r(h) \propto h^x$. This is mainly to simplify my exploration, and to allow my exploration to remain focused on a goal. Further areas of research could include slopes of other functions, such as sinusoidal functions and piecewise functions. Additionally, I only investigated the base of the top, disregarding the handles. Although this limits how generalizable my findings are, its effects could be easily calculated by adding the moment of inertia of the handle into the moment of inertia of the whole top. Lastly, I ignored the complications in precession when angular speed of a top drops below the threshold required to remain a steady precession. This is because at that point, the top starts to wobble and the precession becomes complicated (beyond my understanding of physics). However, it is still a general case where the lower the deceleration, the longer the top spins for. This is also a reason why it will be challenging to calculate the exact duration the top spins for. Realising such complications, I decided to focus my investigation on the relationship between spinning time of the top and its base shape as opposed to determining an exact duration.